DEFENDING GLOUCESTERSHIRE AND BRISTOL

DEFENDING GLOUCESTERSHIRE AND BRISTOL

THE MILITARY LANDSCAPE FROM PREHISTORY TO THE PRESENT

MIKE OSBORNE

FONTHILL

Fonthill Media Language Policy

Fonthill Media publishes in the international English language market. One language edition is published worldwide. As there are minor differences in spelling and presentation, especially with regard to American English and British English, a policy is necessary to define which form of English to use. The Fonthill Policy is to use the form of English native to the author. Mike Osborne was born and educated in England; therefore British English has been adopted in this publication.

Fonthill Media Limited
Fonthill Media LLC
www.fonthill.media
office@fonthillmedia.com

First published in the United Kingdom and the United States of America 2024

British Library Cataloguing in Publication Data:
A catalogue record for this book is available from the British Library

ISBN 978-1-78155-889-8

Typeset in 10pt on 13pt Sabon
Printed and bound in England

Contents

List of Abbreviations

AA	Anti-aircraft
AA	Automobile Association
AAOR	Anti-Aircraft Operations Room
AAP	Aircraft Acceptance Park
ACF	Army Cadet Force
ADGB	Air Defence Great Britain
AFC	Australian Flying Corps
AFDD	Air Force Distribution Depot (fuel)
AFRD	Air Force Reserve Depot (fuel)
AFS	Auxiliary Fire Service
AI	Airborne Interception (Radar)
AM	Air Ministry
AMWD	Air Ministry Works Department
ANZAC	Australia & New Zealand Army Corps
ARC	Army Reserve Centre (formerly TA Centre)
ARD	Aircraft Repair Depot
ARG	Airfield Research Group
ARP	Air Raid Precautions
ARS	Air Raid Shelter
ASE	Admiralty Signals Establishment
ASU	Aircraft Storage Unit (RAF)
A/T	anti-tank
ATA	Air Transport Auxiliary (civilian ferry pilots)
ATC	Air Training Corps
ATS	Auxiliary Territorial Service (1938–49, then WRAC)
BAC	Bristol Aeroplane Company

BAT	Beam Approach Training
BCF	British Concrete Federation
BEF	British Expeditionary Force
BHQ	Battle Headquarters
CBA	Council for British Archaeology
CD	Civil Defence
CCF	Combined Cadet Force (successor to JOTC)
CD	Civil Defence
CFS	Central Flying School
C-in-C	Commander-in-Chief
CO	Commanding Officer
C/Sgt.	Colour-Sergeant
CSM	Company Sergeant-Major
CTA	County Territorial Association
DAMS	Defensively Armed Merchant Ship (First World War)
DEMS	Defensively Equipped Merchant Ship (Second World War)
DFW3	Directorate of Fortifications & Works, Department 3
DH	de Havilland
DL	Defended Locality
DPs	Displaced Persons (post-World War II)
DORA	Defence of the Realm Act (emergency legislation)
EMS	Emergency Medical Service
(E&R)(S)FTS	(Elementary & Reserve)(Service) Flying Training School
FRIBA	Fellow of the Royal Institute of British Architects
GCC	Gloucestershire County Council
GC&CS	Government Code and Cipher School
GCHQ	Government Communications HQ (successor to GC&CS)
GCI	Ground Control Interception (radar system WWII+)
GDA	Gun-Defended Area (AA defences)
GHQ	General Headquarters (GHQ Line, GHQ Reserve etc)
GL	Gun-laying (radar)
GPO	General Post Office
GS	Grammar School
GTS	Glider Training School
GWR	Great Western Railway
Ha	Hectare (2.5 acres)
HAA	Heavy Anti-aircraft
HGCU	Heavy Glider Conversion Unit (pilot training)
HER	Heritage Environment Record (ex-Sites & Monuments Record)
HG	Home Guard
HMT	His Majesty's Troopship
ICI	Imperial Chemical Industries

ITC	Infantry Training Centre
LAA	Light Anti-aircraft
LDV	Local Defence Volunteers, later Home Guard (Second World War)
LG	Landing ground
LMS	London Midland Scottish Railway
MAFF	Ministry of Agriculture, Fisheries & Food (post-Second World War)
MAP	Ministry of Aircraft Production (Second World War)
MI5, MI6 etc	Military Intelligence departments
MOS	Ministry of Supply
MP	Military Police
MRRS	Military Road Route System
MT	Motor Transport
MU	Maintenance Unit (RAF)
NATO	North Atlantic Treaty Organisation
NFS	National Fire Service (Second World War)
NRA	National Rifle Association
OAMCU	Overseas Air Movement Control Unit
ORP	Operational Readiness Platform
OTC	Officer Training Corps (school or university cadet corps)
OTU	Operational Training Unit (RAF Bomber/Training Commands)
(P)AFU	(Pilot) Advanced Flying Unit
POW	Prisoner of War
PWE	Political Warfare Executive
QF	Quick Firing (gun)
QAIMNS	Queen Alexandra's Imperial Military Nursing Service
QM	Quartermaster (logistics organisation and personnel)
RA	Royal Artillery
R & D	Research & Development
RAuxAF	Royal Auxiliary Air Force (until 1957)
RAF	Royal Air Force (from 1 April 1918)
RAFVR	Royal Air Force Volunteer Reserve
RAMC	Royal Army Medical Corps
(R)AOC	(Royal from 1918) Army Ordnance Corps (until 1993)
(R)ASC	(Royal from 1918) Army Service Corps (until 1993)
RE	Royal Engineers
REME	Royal Electrical and Mechanical Engineers (formed 1942)
RFA	Royal Field Artillery
RFC	Royal Flying Corps (up to 31 March 1918)
RFCA	Reserve Forces and Cadets Association (successor to CTAs)
RGA	Royal Garrison Artillery
RGH	Royal Gloucestershire Hussars Yeomanry
RGHQ	Regional Government Headquarters

RHA	Royal Horse Artillery
RHQ	Regimental Headquarters
RLG	Relief Landing Ground
RML	Rifled muzzle loader (Victorian heavy gun)
RN	Royal Navy
RNAS	Royal Naval Air Service (until 31 March 1918)
RNR	Royal Naval Reserve
RNVR	Royal Naval Volunteer Reserve
(R)OC	(Royal from 1941) Observer Corps
RTR	Royal Tank Regiment—mechanized cavalry units
RVC	Rifle Volunteer Corps (1859–81)
SAC	Strategic Air Command (USAF, post-Second World War)
SAM	Surface-to-Air Missile
SAS	Special Air Service
SF	Special Fires (Bombing Decoy—aka 'Starfish')
S/L	searchlight
SLG	satellite landing-ground
SMA	School of Military Aeronautics
SOC	Sector Operations Centre
SOE	Special Operations Executive
SOS	Service of Supply (US Army logistics organisation)
STS	Special Training School (SOE)
TA	Territorial Army (from 1920–39 and 1947–2014)
TAC	Territorial Army Centre (drill hall 1947–2014, see ARC)
tb	temporary brick (single brick with buttresses in RAF buildings)
TDS	Training Depot Station
TF	Territorial Force (from 1908–1918)
TNA	The National Archive (formerly Public Record Office, Kew)
TOB	Temporary Office Building (Government offices from 1940)
u/g	underground
UHF	Ultra High Frequency (radio wavelength)
UKWMO	United Kingdom Warning & Monitoring Organisation
UP	unrotated projectile (see ZAA)
USAAC	United States Army Air Corps (until 1941)
USAAF	United States Army Air Force (Second World War from 1941)
USAF(E)	United States Air Force (Europe) (post-Second World War)
VAD	Voluntary Aid Detachment (First World War)
VB	Volunteer Battalion (volunteer elements of line regiments)
VHF	Very High Frequency (radio wavelength)
VP	Vulnerable Point
VR	Volunteer Regiment (VTC county units)
VTC	Volunteer Training Corps (First World War Home Guard)

WAAC	Women's Auxiliary Army Corps (First World War)
WAAF	Women's Auxiliary Air Force (later WRAF)
WHDC	Women's Home Defence Corps
WLA	Women's Land Army
WRAC	Women's Royal Army Corps (from 1949–92)
W(R)VS	Women's (Royal) Voluntary Service
W/T	Wireless telegraphy
YM/WCA	Young Men's/Women's Christian Association
ZAA	AA rocket batteries

Introduction and Acknowledgements

Each of the previous ten counties which I have covered in my historic military landscape series has been different, and Gloucestershire and Bristol are no less a contrast, consisting of at least three distinct areas: the Cotswolds, the Severn basin, and the Forest of Dean, and including fine examples of almost every type of defensive structure. The River Severn and both Avons have formed significant borders over the centuries, giving Bristol, Gloucester, and Tewkesbury, especially, strategic significance for Britons, Celts, Romans, Anglo-Saxons, Normans, medieval kings and magnates, cavaliers and roundheads, Georgians, Victorians, and so on into the twentieth century. The evidence for this is all around: prehistoric hillforts, Roman forts and settlements, castles of earthwork and masonry, grand fortified houses, forts from the Civil War, Victorian barracks and drill halls, airfields, and anti-invasion defences. A second visible element is the passive defence afforded by ARP, Civil Defence, and electronic surveillance. The region's perceived distance from foreign threat, on the western edge of the country, contributed to its importance in training, munitions production, and logistics. The ordnance factories of the two world wars were prefigured by the medieval weapons industry of the Forest. The area has figured prominently in national developments from the Roman invasion; to the war of succession between Stephen and his cousin; the Wars of the Roses, one of whose decisive battles was fought at Tewkesbury; the English Civil War, during which important sieges were played out twice at Bristol and once at Gloucester with countless skirmishes and occupations by garrisons, and the final Battle of Stow; and the periodic militarisations of the country forced by the threats of invasion and the twentieth century's inexorable move towards total war. Significant and eclectic innovations extend from the production of lead shot, through early flight and aircraft production to women's policing and jet propulsion.

Administratively, Britain appears to have an obsession with re-drawing boundaries and tinkering with local authority jurisdictions, so some definition is necessary. While Bristol traces its existence as a county in its own right to the fourteenth century, its importance as a port in the Severn estuary ties it closely into Gloucestershire, so in all references to that county, Bristol is included. I have therefore, somewhat arbitrarily, decided to discount the brief existence of the county of Avon, taking the present extent of Bristol to define my southern boundary. More specifically, for the purposes of this book, since its main sites are to the east of the county boundary, the airfield at Kemble will be regarded as lying predominantly in Wiltshire, while that at Long Newnton, straddling the Fosse Way, will be deemed as belonging in Gloucestershire.

The process of writing this book, owing to the conditions imposed by the pandemic, has been different to those obtaining for previous volumes. Although Pam and I managed to carry out fieldwork, visiting almost all the sites mentioned in the book, during two week-long stays in Berkeley and numerous daytrips, much of the research for this volume was necessarily carried out during various lockdowns at home, using my own reference materials and the internet. Therefore, there have been fewer opportunities to encounter local historians or to visit local libraries or archives, and the number of those to be gratefully acknowledged for their help is consequently much reduced. I must, however, thank Gordon Williams for information relating to Cirencester and Adrian Armishaw, as ever, for generously sharing illustrations. I must also record our gratitude to Wendy and Stuart at the Malthouse, Berkeley, for the relaxed but safe comfort of their hospitality during difficult times. As always, my special thanks go to my wife, Pam, for her usual help and support, and to Jay, Jamie, Josh, and Alan at Fonthill Media for their friendly professionalism and for continuing to indulge my writing habit.

Mike Osborne
Market Deeping
December 2024

1

Fortifications in Gloucestershire Prior to the Norman Conquest

Prehistoric Fortifications

There are about thirty-five hillforts in Gloucestershire, mainly on the high ground of the Cotswold escarpment on the eastern side of the county but with several more lying across the Severn in the Forest of Dean and a few in the south of the county around Bristol. Many of the early hillforts, promontory forts, cross-ridge dykes, and forts defended by complete circuits of ramparts such as Crickley Hill and Leckhampton, appear to have been abandoned by about 400 BC. The final developments of the hillfort are characterised by multiple lines of defences, and in Gloucestershire, these forts lay within the lands of the Dobunni tribe. Some hillforts such as Uley Bury, Salmonsbury, Sodbury, and Kimsbury did continue in occupation from 300 BC to AD 100, and Welshbury in the Forest of Dean may also belong with this group. Some earlier forts such as Brackenbury Ditches and Cleeve Cloud may have been upgraded at this time by the application of additional lines of defence, or as at Willersey by cutting a 10-foot-deep (3-metre-deep) rock-cut ditch to isolate a smaller area within the larger original fort. But it appears to be the case that defences *per se* were subordinated to a projection of strength and power, reflected in the sheer effort expended in their construction as some powerful local authority was clearly involved in marshalling the requisite resources. There is still considerable mystery attached to the identification of many of these prehistoric sites. Drakestone Point, Stinchcombe, often described as the best viewpoint in Gloucestershire, has been interpreted, among other things, as an Iron Age camp, a Roman signal station, and an eleventh-century motte. A survey carried out by the GCC Archaeology Service in 2005 revealed that four lines of manmade banks and ditches defended the three terraces on the end of the point but was hesitant about being more specific about date. Other sites at

South Dowdeswell, Birdlip, Charlton Kings, and Dixton Hill have been rejected outright or seen as dubious. Despite the lack of a consensus on the purpose of such structures, many have clearly been built with defensive features. It cannot now be known whether these were to protect the occupants on a daily basis; to provide secure storage for food-stocks; or to provide an emergency refuge in times of conflict. Some sites such as Soudley Camp, where a large corn-grinding quern has been found, have revealed indications of domestic occupation, its presence pointing away from industrial use, usually characterised, in that particular locality, by iron-working waste. Parts of a quern were found at Sodbury and excavations at Windrush Camp found traces of a scattering of round huts and storage pits and the Rings on Cleeve Hill may offer an example of hut platforms. The early phase of building at Crickley Hill included timber halls, 24 feet (7.4 metres) wide and anything up to 80 feet (25 metres) long, of which at least two had central hearths, all flanking a roadway, within an enclosure surrounded by a timber palisade strengthened by drystone walling.

Some sites have revealed signs of actual conflict, one indication of which may be the fire-reddened stones, visible in several places at Beckbury Camp and at Leckhampton Hill, around the fort's entrance and in the outer ditch, or the slingshot recovered at Crickley Hill and the evidence of burned buildings. Some sites are thought to have been involved in recorded military actions. Symonds Yat fort, placed on a small triangular promontory and defended by ramparts only on the south side, is reputed to have been the scene of an encounter between the invading Romans and Caratacus, leader of the Catuvellauni who was forced to make his escape across the Wye to the fort on the Doward, continuing the fight

Old Sodbury hillfort, the view through the double rampart and ditch from inside the fort.

for a further twenty-seven years. Welshbury Camp is thought to have been the site at which those of the Dobunni tribe who had chosen to resist the Romans made their final stand. A recent re-examination of Iron Age warfare suggests that the range of slingshot would have been taken into account when selecting a site and laying out the defences. An early phase of the Crickley Hill settlement, dating back to *c.* 3500 BC, appears to have culminated in conflict confirmed by the presence of hundreds of flint arrowheads found close to burned out gateways. An examination of the later defences of Crickley Hill concluded that the fort was vulnerable to attackers firing slingshot but points out that the fort remained unfinished and additional outer defences might have been envisaged but never built before this fort was overrun and abandoned by *c.* 400 BC.

Based on their specific locations, prehistoric hillforts generally fall into several distinct categories. Several of the forts in this region can be assigned to the category described as promontory forts, which often occupied spurs where, with steep scarps on three sides, only a simple cross-ditch was needed to secure the site. Brackenbury Ditches encloses 4 acres (1.5 ha). at the tip of a spur with a rampart and ditch. On the north-east side, an outer rampart stretches across the neck of the spur forming an outer enclosure. Camp Hill at Lydney is a typical example of such a fort with steep natural defences on the east, south and west sides, and while only low banks follow the contours of the hill to defend three sides, the north side is defended by double banks and ditches up to 7 feet (2 metres) deep. At Ring Hill Camp, Haresfield, banks enclose 10 acres (4 ha) at the end of a spur. The banks follow the edge of the scarp and are pierced by at least three openings approached via terraces or hollow ways. Around 500 yards to the east, the Bulwarks cuts off a wider neck of the same promontory. At Cross-ridge Dyke on Juniper Hill, Edgeworth, a short dyke cuts off the end of a spur. Around the Stroud area, simple cross dykes such as those found on Coopers Hill, Brockworth; Pen Hill, King's Stanley; and Standish Woods, Randwick, were dug to provide protection. Not all promontory forts were so compact. Nottingham Hill is a large promontory fort of 120 acres (48 ha) with a double rampart and ditch on its south-east side with the scarp of the hill defending the other three sides.

Where it was not possible to cut off a spur with a single defensive rampart and ditch, a semi-circular enclosure might be formed up against a steep scarp forming its straight side. Hinton Hill Camp, Dyrham, is a small promontory fort whose single rampart and ditch curves in an arc to enclose an area of around 12 acres (5 ha) and at Ablington Camp, Bibury, a semi-circular single bank and ditch encloses around 8 acres (3 ha) on the edge of a plateau above the River Coln. Norbury Camp, Colesbourne, occupied a spur and its oval single bank and ditch enclosed about 8 acres (3 ha). Bloody Acre Camp, Cromhall, sits on the summit of a spur above a stream. Here, an irregularly shaped area of 14 acres (5.5 ha) is enclosed, with a double bank and ditch on the north, triple banks and ditches on the south and west, but only the steep natural scarp on the east. Beckbury Camp lies on the edge of a plateau and an area of 5 acres (2 ha) is enclosed on two sides by a stone

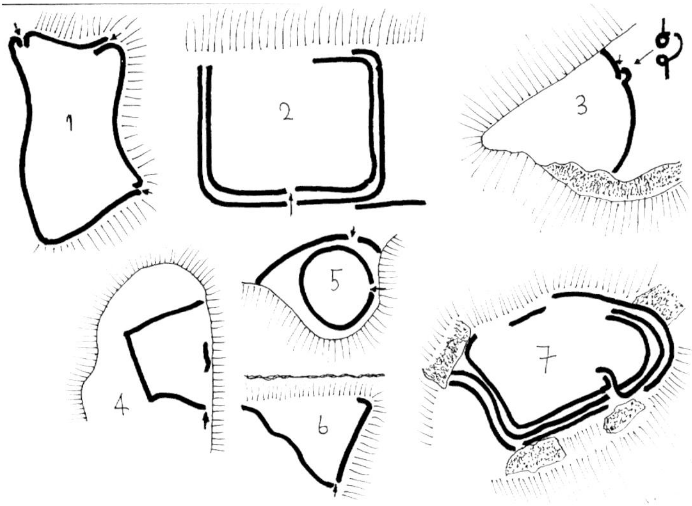

Sketch plans of Gloucestershire hillforts: 1 Uley Bury, contour fort; 2 Old Sodbury, contour fort; 3 Crickley Hill, promontory fort; 4 Upper Dowdeswell, promontory fort; 5 Brackenbury Ditches, promontory fort; 6 Ablington, promontory fort; and 7 Painswick, contour fort.

rampart which formerly had an exterior ditch. The other two sides are protected by the scarp. Sites were often chosen for such strengths and Dean Camp, Coln St Aldwyn, around 12 acres (5 ha) in area, stands on a spur above the confluence of two streams. It has just a bank on two sides and relies on the natural scarp on the other two. Lancaut is a bivallate promontory fort of roughly triangular shape, parts of which were later integrated into Offa's Dyke, facing east and appearing to block the narrow neck which gives access to land contained within a loop of the Wye, measuring 2.5 acres (1 ha). Two further camps lie to the south, on the opposite, Welsh, bank of the Wye, also exploiting the natural strength of these loops of the river. A similarly strategic location can be found at Clifton Down Camp which stands above the precipitous cliff of the Avon Gorge. It is defended by a semi-circular double rampart and ditch which seals off a roughly oval area. The cliff's edge requires no further defences, but a path descends to a spring below and on down to the ford below the present suspension bridge. On the opposite, Somerset, side of the gorge stand two similar camps which may have belonged to an opposing tribe or have been part of a complex defending the river crossing.

Although many hillforts make use of spurs or scarps for their defences, only providing serious barriers on their exposed flanks, others, described as contour forts occupy whole hilltops, often in easily defended locations, where all-round defences are necessary. These can be very small as is Shenberrow, a D-shaped hillfort, enclosing just 2.5 acres (1 ha) with a single bank and ditch on the curved side and a slight bank along the straight which follows the scarp. Bury Hill Camp on Winterbourne Down is another small multi-vallate hillfort covering just 5 acres (2 ha) standing on a plateau overlooking the River Frome. The two circuits of banks and ditches are closely spaced and there are entrances on the north-east, north-west, and south-east. That on the north-west has an outer bank as an added defence not found at the other two. Towbury Hill Camp consists of a slight single rampart and ditch enclosing a roughly triangular area of 10 acres (4 ha). It stands on a low-lying plateau almost equidistant from the Severn and the Avon which are less than 2.5 miles (4 km) apart at this point and overlooks a brook. Significantly larger is North Dowdeswell, a univallate hillfort occupying 14 acres (5.5 ha) on the eastern side of a spur. The rampart survives on the south and west sides but appears to have slipped away down the steep scarp on the east, with an entrance which may have been at the south-east angle at the top of the slope. Little Sodbury is a roughly rectangular enclosure covering 24 acres (9.5 ha) with a double rampart and ditch on three sides but only a single one on the west and entrances at the north-west and south-west corners and in the middle of the east side. Uley Bury Camp is defended by a continuous rampart enclosing 38 acres (15 ha). There are entrances at three corners, two of them formed by ramparts, in-turned to create gate-passages, and the third by overlapping them. All three were formerly protected by exterior banks and ditches. Camp Hill, Rockhampton, sometimes referred to as Thornbury, is a multi-vallate hillfort occupying the summit of a prominent hill and enclosing an area of 15 acres (6 ha), defended by double rampart and ditch, clearly visible on the north, east and south sides.

Uley Bury, one of three entrances penetrating the rampart.

A further category is the defended settlement, a class of monument less common than the hillfort, and built in less promising locations. One example is Elberton Camp, a small, defended settlement in the shape of an irregular rectangle, covering only 3.75 acres (1.5 ha) and defended on the north by a steep natural slope, and on the other three sides, by a single stone-built bank and ditch. Another is at Batsford, Moreton-in-Marsh, adjacent to Queen Victoria's Garden, where an Iron Age defended settlement of roughly rectangular form measuring 200 by 210 feet (65 metres by 60 metres) is enclosed by a rampart and external ditch. Another of these low-lying camps is Oldbury Camp or The Toot(s) at Oldbury-on-Severn, occupying a strategic, low-level site overlooking a stream which flows into the Severn just over half a mile (1 km) to the west. There are double banks and ditches on the north and east and a single bank on the west. The south side now consists of an earthwork platform which has been interpreted as a wharf, reached at high tide, but may be nothing more than the result of agricultural activity. Salmonsbury Camp, Bourton-on-the-Water, is an Iron Age fortified settlement of 56 acres (22 ha) also occupying a relatively low-lying site on a gravel platform between two rivers. Double banks and ditches survive on the north-east and south-east sides and there are traces of an annexe at the eastern angle. Both Salmonsbury and Oldbury have been included in the category of valley forts, or marsh forts, occupying low-lying, but strategically significant, locations, often close to the confluence of waterways, and employing the same techniques of fortification as their elevated cousins but substituting the benefits of water for those of height.

More substantial, but less coherent than these discrete settlements are the possible tribal centres of the Dobunni marked by sequences of dykes enclosing large areas. A series of nine dykes associated with an Iron Age and early Romano-British settlement at Bagendon has been tentatively identified as the possible site of *Corinion*, as described in Ptolemy's itinerary, as the Dobunnic *capitas* or *oppidum*, which preceded the Roman town of *Corinium* or Cirencester, 3 miles (5 km) to its south. The discontinuous lengths of dyke enclose an area variously defined as from around 200–500 acres (80–200 ha). Some 4 miles (6 km) south-east of Gloucester, a series of earthworks possibly constituting a link between High Brotheridge Camp and Coopers Hill appear to form another large, *oppidum* of the Dobunni covering 200 acres (80 ha). Evidence of stone revetments to banks at High Brotheridge have been noted as well as deep rock-cut ditches. A third site, a series of dykes on Minchinhampton Common, has been suggested as another possible *oppidum* site but has been more recently interpreted as a post-Roman defence against encroaching Saxon settlers and/or the reuse of earlier dykes as medieval park boundaries.

Although the majority of Iron Age forts were sited in naturally defensible locations, strenuous efforts were made to enhance the natural strengths afforded by steep scarps. While a single low rampart might suffice on the edge of a precipitous drop, where the approach was gentler, anything up to four banks and ditches might be dug. Painswick Beacon or Kimsbury Fort is defended by triple ramparts and ditches. Bloody Acre Camp, Cromhall, has double banks and ditches on one side but triple ones on the two more vulnerable sides. Where there were double circuits, the inner ramparts usually overtopped the outer ones. A ditch dug to a depth of 7 or 8 feet (2 or 2.5 metres) could produce a rampart which stood twice that height from the bottom of the ditch and would represent a significant obstacle. The inner bank at Camp Hill, Rockhampton, rises to 11 feet (3.5 metres) while the outer one is lower. These heights would be further increased by the timber palisade which usually crowned the ramparts. This would often be quite robust and carry a fighting platform which would be carried over gateways, sometimes through a tower straddling the gate passage, which would be sealed with double wooden gates at each end. Excavations in 1926 at Winterbourne Down found that the ramparts were formed of a masonry core of large flags topped by smaller stones and soil, all supported by a revetment of smaller flags. It was remarked at the time that construction was unhurried and that the fine rubble was laid carefully and unhurriedly by the basketful with no evidence for a pressured build under conflict conditions. Despite fine preservation in the relevant areas, no evidence of the postholes of a palisade was found.

As with most fortifications, it was the entrances which received most attention and a variety of techniques were employed to strengthen them. One method was to restrict access as at Brackenbury Ditches, where the entrance in the south side is approached along a ledge accessed by a break in the outer rampart, placing those entering in full view of the defenders and at a lower level, the intention being to create killing grounds. The inner enclosure at Ring Hill Camp, Haresfield, is pierced by at least three openings approached via terraces or hollow ways and around 500 yards to the east, the Bulwarks, cutting off a wider neck of the same promontory, has four entrances approached across causeways in the ditch. The entrance at Norbury Camp, Colesbourne, lies between overlapping ramparts at the southern end of the circuit furthest away from the neck of the promontory. Uley Bury Camp has entrances at three corners, two of them formed by turning the ramparts inwards to create gate passages, and the third by overlapping them. All three were formerly protected by exterior banks and ditches and approached up steep slopes and one of these entrance passages has been found to have been lined with turves and timber. At Painswick Beacon, the inner rampart at the entrance near the north-east angle is extended into the interior of the fort to create a hooked passage overlooked from the rampart walk. Bury Hill Camp on Winterbourne Down is defended by two closely spaced circuits of banks and ditches with entrances on the north-east, north-west, and south-east. That on the north-west, presumably judged to be more vulnerable, has an outer bank as an added defence not found at the other two. Leckhampton's entrance was flanked by semi-circular guard chambers. Following a period of abandonment, Crickley Hill fort was reoccupied and refortified by a different tribe or clan. Excavations have revealed stone bastions flanking the inner entrance, with an outwork accessed through an outer gateway, and remodelled ramparts. Somewhat surprisingly, at Sodbury the entrances through the two ramparts line up rather than, as might be expected, being staggered.

Depending on the availability of materials, Iron Age forts were constructed using a variety of techniques. At Leckhampton, the rampart was 20 feet (6 metres) wide and 6 feet (1.8 metres) high with a stone core revetted by drystone walls tied together at the base with timbers laid crossways and excavations at Crickley Hill have revealed a similarly timber-laced rampart with stone revetments on inner and outer faces. The rampart of the short cross-ridge dyke, which cuts off the end of a spur on Juniper Hill, Edgeworth, has been excavated and shown to be mainly rubble with, on its outer face, a 3-foot-high (95-cm-high) drystone revetment. Excavations at Welshbury Camp have revealed that the ramparts were constructed of large angular stone blocks laid on the bedrock itself or onto surface rubble. Burgh Walls at Clifton has been suggested as an example of the technique of vitrification. In this process, the unbonded loose stones of the rampart are mixed in with lime and brushwood. Once the brushwood is fired, and water poured on, the mass is left to smoulder in the rain, and turns into concrete, forming a solid mass as the core of the rampart. More earth and stones may be added to this firm foundation. It must be noted that this explanation has been challenged with other causes including deliberate destruction by fire as a possible alternative.

Cross-section of a hillfort rampart and ditch. Timber uprights supported by dry-stone walls form a framework for the boxes which contain packed earth and rubble. An internal rampart of earth supports the fighting platform made of logs behind the palisade. Along the outside of the ditch runs the counterscarp, a low rampart which both impedes the assault and allows defenders to fire down on attackers as they approach and as they crest the bank.

There was no set lifespan for the hillforts. Sometimes, new sites were chosen for a fresh settlement, as at King's Weston Hill Camp at Henbury, which appears to have been abandoned by the time neighbouring Blaise Castle Camp came into use. Destruction was sometimes followed by reconstruction. The first fort at Crickley Hill was captured and the ramparts destroyed by burning. A new fort remained unfinished and had been over-run and abandoned by *c.* 400 BC, but the site showed signs of occupation again into the Roman era. Salmonsbury reveals evidence of earlier occupation suggesting that the defences were only constructed a short while prior to the Roman invasion, but the fort continued in occupation into the Roman era. Finds included silver Dubonnic coins and a hoard of currency bars. Roman coins have been recovered from a number of other hillforts including Sodbury. A small camp near Woodmancote, North Cerney, appears to have been largely destroyed when a Romano-British settlement was superimposed over it. Other sites have disappeared over the centuries and continue to succumb to quarrying, the construction of new houses and roads.

Roman Fortifications in Gloucestershire

For a short while in the decades following the first-century AD Roman conquest of Britannia, the county was important militarily. Gloucestershire was traversed by two major Roman roads: the Fosse Way ran from Exeter to Lincoln, south to north through Cirencester where it crossed the Ermin Way which ran east to west from Silchester to Gloucester and beyond. Parallel to the Fosse Way, a further Roman road ran from the fort at Sea Mills (*Abona*) to Gloucester. In an introductory essay in Verey (1), the area bounded by these two roads is described as a militarized zone, but this situation could only have obtained during the short period leading to the shift of military attention to South Wales, long before the end of the century. Throughout the four centuries of Roman occupation, activity was centred on the two major urban centres of Gloucester and Cirencester with further minor centres in the county, some of which enjoyed only temporary significance.

Gloucester (*Glevum*)

In AD 49, in the build up to the Roman conquest of Wales, the XXth Victrix Legion was deployed from Colchester to Kingsholm to establish a fort there as an operational base, and to secure the Severn ford. This earliest fort would have been the usual playing card shape with earthen ramparts topped by a timber palisade and surrounded by one or more V-section ditches. The rampart was built on foundations fronted by a deep turf wall and was 37 feet (11.5 metres) wide. The four gates, one in the centre of each side, might each have been overlooked by a timber tower, possibly built over the gate passage, with the rampart walk being

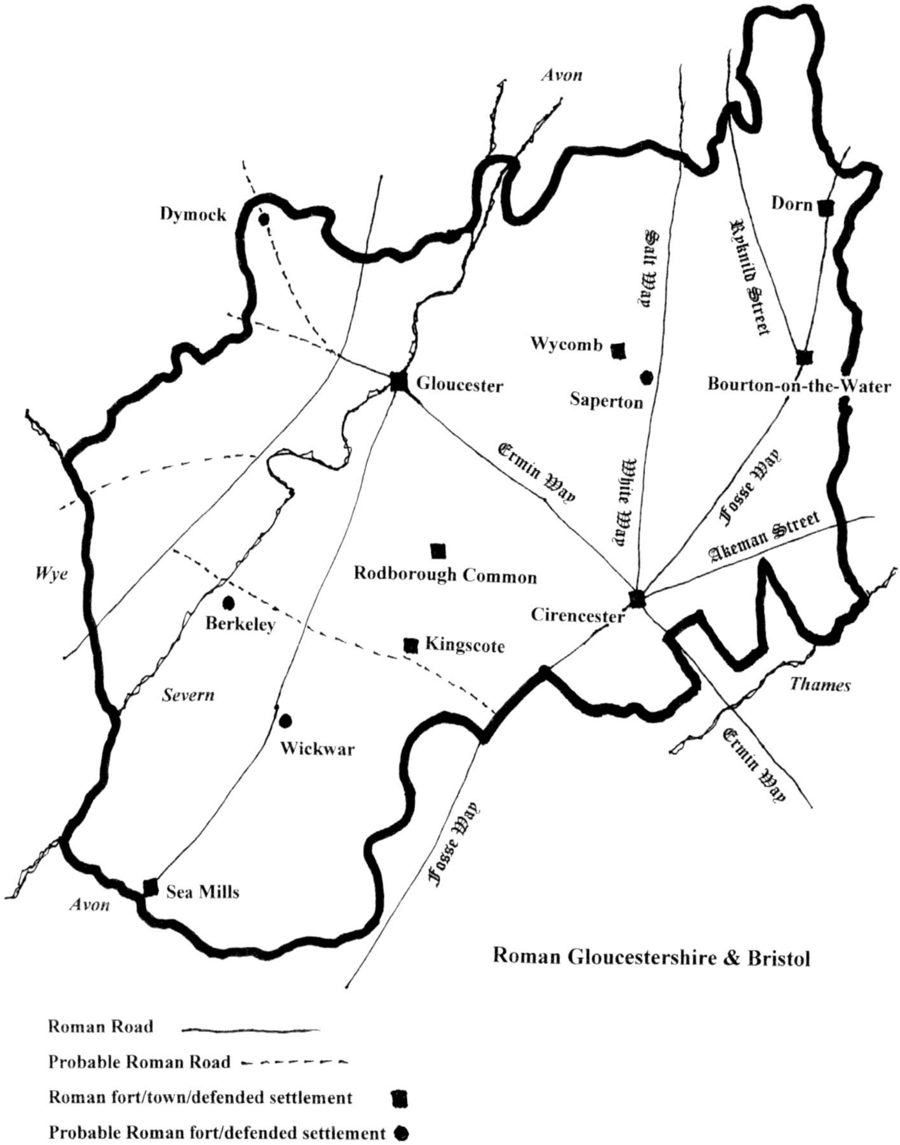

Map of Roman Gloucestershire showing roads and forts mentioned in the text.

carried across it. Inside the walls, timber barracks, storerooms, granary, stables, and workshops would have occupied the four quadrants with a headquarters building in the middle where the four main streets crossed. Probably as a result of the area being prone to flooding, a new site was chosen for a larger legionary fortress in AD 64. This new fort, which stood near to the present Gloucester city centre, was both stronger and more extensive, covering an area of 43 acres (17 ha) alongside the river. The ramparts were over 10 feet (3.5 metres) in height and made of packed clay, held in place by timber lacing, and faced with stacked turves, inside the usual double V-shaped ditches. A timber tower stood at each corner; the gates were flanked by paired towers; and there may have been interval towers as well, astride the rampart. The plan and general layout were both similar to those of the earlier fort but some of the buildings were constructed of plastered clay on a timber framework. The floor of the granary stood on a grid of beams supported on low stone pillars, and some of the barrack huts rested on dwarf stone walls. Having been involved in the conquest of South Wales and the establishment of a legionary fortress at Caerleon/*Isca*, the XXth Legion went on to take part in campaigns in the north, sometime in the AD 70s, but returned to Gloucester, giving its fortress a further makeover. A stone facing of large un-mortared limestone blocks was added to the clay ramparts; stone towers were added to the four gates and the opportunity was also taken to scour or recut the ditches. The North Gate was L-shaped with the face of the gate shielding a dog-leg gate passage and has been dated to AD 96. The East Gate was a square tower with a lateral wall, with its front face flush with the city wall. It was of similar date to the North Gate but had much heavier foundations. At most a decade after this work had been completed, but possibly even sooner, Gloucester had lost its military importance and became, as did other similar former legionary bases, a *colonia* for retired soldiers and their families. Gradually, military buildings were replaced by private houses, baths and the other buildings of a civil settlement. The HQ building became the forum, and dwellings spread outside the walls to create suburbs. Although the town had outgrown its defensive perimeter, the walls were not neglected. Their depth was increased by the enhancement of the inner earth bank supporting the wall walk which had probably suffered from erosion by the end of the second century. Stone corner towers were built as well as interval towers between them and the gates. As the third century progressed, raids from north-west Europe and Ireland increased in frequency and many of the towns of Roman Britain found it necessary to improve their defences. At Gloucester, vulnerable to ship-borne raiders sailing up the Severn, the walls were subject to a radical rebuild. The massive stone facing blocks now became the new, and very substantial, foundations for a higher wall of massed concrete faced with stone. Those sections of wall being upgraded in the later stages of this process saw the foundation blocks bedded onto a layer of rubble and mortar on top of timber piles sunk deep in natural sand. One particular development seen at many of the towns having their walls strengthened at this time, was the construction of external towers designed to carry artillery—mainly the *ballista*, a development of an earlier stone-throwing catapult, which fired bolts or spear-sized arrows, or the *onager* which propelled rocks. Building the towers in advance of the rampart allowed the artillery to enfilade the walls and to push attackers further back. Ditches remained in use as effective obstacles to an assault. Defenders in the interval towers and on the wall-head might have been equipped with the smaller bolt-firing *manuballista*, still mounted on a frame, but operated by a single individual. The massive blocks of stone used in the construction may be seen at several points on the circuit of the city's Roman walls, notably in the Eastgate Chamber and near the South Gate. The Roman East Gate comprised two square turrets flanking an internal gate passage.

Sea Mills (*Portus Abona*)

This settlement stood at the confluence of the rivers Avon and Trym and was established in the first century as a military post, with a civil settlement following shortly afterwards. The relatively early date of *c.* AD 46–48 has been suggested with the port providing a base for Roman warships and a supply base for the legions converging on Kingsholm. *Abona* then acted as the supply base for the operations led by Frontinus against the Silures of South Wales in AD 74–78. After the XXth Legion had moved on, there appears to have remained a small military presence at *Abona* in order to protect such an important supply point. Evidence for masonry having collapsed into a ditch, points to the previous existence of town walls, something apparently confirmed by later descriptions into Anglo-Saxon times. These same charters of 969 and 984 also refer to a watchtower of supposed Roman provenance standing on a site around 320 feet (98 metres) above sea level, in the south-east corner of the assumed ramparts, with views along the Avon. It is still uncertain to what extent *Abona* remained a base for the *Classis Britannica* given the existence of the legionary base at Caerleon in Gwent. Later, into the third century, when raiders from Ireland were threatening to disrupt Roman rule, the centre of fleet operations may have been shifted to the shore fort at Cardiff. Raids continued throughout the fourth century, the danger peaking in 367–68 when attacks on Roman Britain came from all sides, severely testing the ability of the fleet to cope. Evidence would suggest that villas in the region suffered from these incursions even if places like the defended towns survived. The port will have handled commercial traffic throughout its existence and occupation of the township and port continued into the fourth century when it appears to have been abandoned. Excavations have uncovered only the footings of buildings thought to be shops and industrial premises such as potteries, now preserved and visible, but finds have been attributed to military occupation.

Cirencester *(Corinium Dobunnorum)*

This important town was situated at the strategic road junction and river crossing a short distance from the Dubonni *oppidum* of Bagendon. The line of the Fosse Way, from about AD 47, marked the extent of conquest for a while and the Roman invaders established a fort here in the early years. This, it has been suggested by Neil Holbrook, may have been more of a gesture of support for the generally friendly Dubonni, rather than a symbol of occupation and oppression. The Dubonni appear to have opted out of the bitter and bloody, but ultimately unsuccessful, resistance to Rome of their neighbours in the south-west. The exact location and extent of the earlier fort, or marching camp, is not entirely clear, and conjecture has centred on its relation to the alignment of the Fosse Way. This camp was superseded by a second fort, referred to as the 'Leaholme fort', constructed soon after AD 50 as the planned layout crystalised. The fort was garrisoned by successive units of auxiliary cavalry who would move on in AD 75 once the military focus had shifted to the west and north. It was built of timber with turf ramparts and two outer ditches. This was demolished as the town grew, initially also surrounded by earth banks and a palisade. The town was walled in stone in the late second or early third century when it has been suggested that the rampart and stone wall with its stone gates and interval towers were contemporaneous rather than, as at many Roman towns, an incremental development of their defences. The new stone walls, set into the forward face of the earthen rampart, were 20 feet (6 metres) high and had interval towers placed along the irregular course of the walls. The Verulamium Gate has been shown to have been an impressive construction with a pair of three-storey D-shaped towers flanking double carriage arches with a pedestrian passage to each side, entered across the defensive ditch over a stone bridge. It is quite likely, befitting the importance of the town, that the town's other three gates would have been equally monumental. The town covered 240 acres (96 ha) and was the second largest Roman town in Britain with a population estimated at between 10,000 and 12,000 people with its walls measuring nearly 2 miles (3 km) around. As happened in Gloucester, polygonal projecting bastions were added later in the fourth century. The town was laid out on a grid system with the important civic and market buildings in the centre where the four main streets crossed, and a large amphitheatre outside the walls. It is likely that *Corinium* served as a provincial capital into the fourth century, the town's significance in the country's infrastructure being underlined by its location at the conjunction of four important roads: the Fosse Way, Ermin Way, Akeman Street, and White Way which runs due north becoming the Salt Way around 12 miles (20 km) out of town. Fragments of Cirencester's walls, including the base of a projecting bastion, may still be seen in the abbey grounds.

Cirencester, a stretch of Roman town wall and the base of a bastion.

Roman Minor Towns and other Roman Military Sites

Across the country there are many Roman settlements which grew up following the establishment of a fort or a road station and several of Gloucestershire's Roman settlements may have originated in this way, particularly those which lie close to major arteries such as the Fosse Way or other such roads. Hall End (a possible *Milidunum*) at Wickwar is a small, possibly defended, Roman town lying between Sea Mills and Gloucester, on the line of the north–south Roman road. The site contains traces of substantial buildings of both timber and stone. A possible identification of a ditch feature suggests that it may represent the corner of a Roman fort implying that the town either occupied the interior of a former fort or was given its own defensive ditch subsequently with an entrance gate also mentioned in reports. At Dorn is a rectangular enclosure with massive earthwork banks lying alongside the Fosse Way near Moreton-in-Marsh. The town was divided into six *insulae* and streets led toward what appear to be gates onto the Fosse Way. It has been described as unlike any purely civil settlement so may have had some official or administrative, but not necessarily military, significance. Finds, including Samian ware, indicated occupation from the first to the fifth centuries. Bourton-on-the-Water, where Ryknild Street branched off to Lichfield and ultimately Chesterfield, has been suggested as the probable site of a Roman fort on the Fosse Way midway between Cirencester and Dorn at the ford which crossed the River Windrush. At Dorn, a small town grew up from the end of the first century, serving travellers with a *mansio* for official travellers and catering outlets. It was of rectangular shape and surrounded by banks and ditches enclosing six *insulae*, and there is a slight indication of the site of a gateway. There were also industrial sites and the settlement continued beyond the end of the Roman occupation when a road ran alongside the river to Salmonsbury Camp. Despite the fortified nature of the site, no overtly military finds have been unearthed. Kingscote is a Roman site, traditionally known as the Chessalls, a name associated with a fortified settlement, and close to a possible Roman road which links the Fosse Way to the Severn. The small Roman town of Wycomb on the north-eastern outskirts of Andoversford may also have begun life as a military station, implanted on an Iron Age ritual site, although no specific evidence for this has yet emerged. Traces of buildings, one of them a basilica-like structure, paved roads, and ditches yielding Roman coins have all been found. It lies close to a Roman road known as the White Way/Salt Way running north from Cirencester. The nearby Pen Hill camp at Salperton, alongside the Salt Way, has also been put forward as a Roman station. The position of Berkeley on the Severn opposite the Roman settlement of Lydney has been proposed as the location for a fort which would have guarded the road from Cirencester to the ferry crossing. Dymock, accepted as lying on the military road running westwards from Gloucester, at its junction with the road to Stretton, each a day's march distant, has been proposed as the likely location for a military presence and its dominant position on the ridge above the river and the speculative existence of a *mansio* may indicate some such official standing, and finds of early coinage and sophisticated pottery point to a military connection—if not an actual garrison then a place frequented by soldiery. It has been put forward as the unidentified settlement of *Macatonium* mentioned in the *Ravenna Cosmography*. However, evidence for this supposition goes no further than ditches and post-holes which may just as well indicate civilian activity as military.

The earthworks at Pen-Y-Bryn, Rodborough Common, previously identified as associated with the *oppidum* at Minchinhampton, have subsequently, on the basis of aerial photographs and finds dated to the first century, been identified as Roman. A V-shaped ditch, 250 feet long, appears to be associated with a military settlement and excavations have produced finds which tend to confirm a first-century military occupation. The prehistoric camp at Oldbury-upon-Severn appears to have been reused in Roman times then put to agricultural use in medieval times, the shape of the earthworks dictating the shape of the later village. A short distance to the south, St Arilda's Church lies in a second earthwork at Cowhill. Although it was unusual for a Roman villa site to be fortified, Chesters villa, dating from no earlier than AD 300 and lying between Woolaston and Lydney on the west bank of the Severn appears to be surrounded by a wall, entered through a gateway, and with a substantial tower at its south-east corner. This is thought to have acted as a lighthouse to aid the passage of river traffic past rocks and into the harbour at nearby Lay Pill. Other locations of navigational beacons from the Roman period have been identified along the Severn. Lydney's hillfort was reused by the Romans but not for military purposes and Cleeve hillfort was occupied in Roman times, as may also have been Blaise Castle camp and Haresfield, Ring Hill, Roman camp.

West of the Severn in the area of the Forest of Dean, five small rectangular earthwork enclosures, measuring up to 100 by 115 feet (30 by 40 metres), with banks and external ditches, have been detected by LiDAR. These have tentatively been identified as Roman fortlets set up to support the westward movement of the legions. Excavation found the sort of pottery associated with Roman military sites of the first century AD and established that they had been abandoned and deliberately slighted around a hundred years later. As the landscape would, at that time, have been more open, they might well have occupied vantage points for the surveillance of newly occupied territory. Although the bank had been substantially eroded the ditch was found to have measured almost 6 feet (1.8 metres) in depth. A small promontory fort at Tytherington has been claimed as the site of a Roman fort but no traces of Roman refortification or occupation have been found in the hillfort. Other likely rejects include Hempstead Camp which has been described as a Roman camp defending the approaches to Gloucester, but although there is evidence of Romano-British settlement of the second to fourth centuries, the earthwork remains date from the medieval period. Supposed by its playing card shape to be a Roman camp, the earthworks at Bitton have since been found to be medieval.

Gloucestershire in Anglo-Saxon Times

In an article in *British Archaeology*, Stephen Yeates, using the Gloucestershire area as his example, suggests that the English landscape owes more to continuity than it does to change. Not only did development arise from Roman estates, such as Withington, but also harked back much earlier to the Dobunni. Taking Gloucester and Winchcombe as his main focal points, he shows how the same settlements existed in Iron Age, Romano-British, and Anglo-Saxon eras, centred along the river valleys, on hillforts, villas, and, ultimately, minster churches, such as that at Bibury, perpetuating the lives of the groups which had always inhabited those places. Max Adams cites Ablington as an example of the sequence from hillfort to villa to early medieval settlement, with the nearby villa site at Barnsley offering a further possible demonstration of such continuity of occupation. Recent suggestions that the Anglo-Saxons practised landscape planning based on a short perch grid (15 feet or 4.75 metres) cited Hempstead and other Gloucestershire locations as examples.

For nearly 200 years after the Romans left, Gloucester with its still formidable fortifications is likely to have provided a base for a local chieftain or petty king. The Roman amphitheatre at Cirencester has been suggested as a possible site for a Romano-British fortified site in the years after the Romans left. Other Roman amphitheatres across the empire were similarly adapted so that possibility is quite plausible. Crickley Hill Iron Age hillfort was apparently reoccupied and refortified during the fifth century when timber buildings, including a granary, were constructed within a palisaded enclosure inside the hillfort. It is possible that the hillfort at Blaise Castle was similarly revived, just as Roman forts on the northern border were occupied by local warlords. Minchinhampton Common at 650 feet (200 metres) above sea level is crossed by the Bulwarks, a bank and ditch which extends for over a mile to the north of the present town. Other lengths, the Rodborough and Amberley earthworks, some of which may be extensions of the main bank, continue the encirclement. The Bulwarks consist of a curving, 1.375-mile-long (2.2-km-long) limestone bank with a stone revetment wall, a berm, and a ditch. It has usually been interpreted as part of the defences of an *oppidum* of the Dobunni, perhaps held against the Romans. However, as it would appear that the *Dobunni* recognised the advantages of Roman rule and raised little resistance, any fortified *oppidum* was unlikely to have seen action. Such a structure, however, does suggest a serious fortification and it has been suggested that the bank may represent a boundary between a British enclave and the lands settled by Saxon incursions in the fifth or sixth centuries. Lacking archaeological artefacts, placenames have been cited in support of this theory, showing areas of Saxon settlement and others where the British tradition may have lived on. According to the *Anglo-Saxon Chronicle*, Gloucester and Cirencester remained centres of local administration into the last quarter of the sixth century, ruled by kings with British or Roman names. All this resistance came to an end with the Saxon victory of AD 577 at Dyrham where three local kings were killed, one being Conmail of *Glevum*, and so began the Saxon colonisation of the area. Michael Wood contends that while the area had now fallen under Anglo-Saxon rule, the majority of the population would have remained British throughout the seventh century and cites a series of Old English charters to assert the continuing survival of social structure and sources of revenue. In 679, St Peter's church in Gloucester was endowed by King Osric with the tribute of three Gloucestershire hundreds: later Berkeley, Whitstone, and King's Barton, suggesting the possibility of an inherited system of taxation.

Queen Ethelfleda of Mercia founded several *burhs* including the unidentified Bremesbyrig. Its location has been suggested as Bromsberrow in the Forest of Dean, but no supporting evidence has been discovered. In Gloucester, St Oswald's minster church, lay outside the city walls on the site of a Roman cemetery by the Severn. Also founded by Queen Ethelfleda of Mercia much further outside the walled area was Kingsholm Palace, which may have hosted the council held in Gloucester in 896. Her nephew, Athelstan, held a council of allied leaders, in effect, a national assembly there, during what has been described as 'the imperial phase of English kingship' and died there in 941. It was still in use in 1051 when King Edward the Confessor convened a meeting of his leading subjects there and assembled troops to put down a revolt, and at Christmas 1085, William I consulted his advisers on what would become Domesday Book. It continued as a royal residence into the reign of Henry III in 1233. The remains of what appears to be an Anglo-Saxon hall analogous to that at Cheddar have been traced and the ruins of a stone house survived into the late 1700s. The palace, on a site now marked by Kingsholm Square, appears to have over-lapped the earliest Roman fort and its chapel may have been built on Roman foundations. Gloucester passed into Mercian possession early in the seventh century and in 909 the remains of St Oswald were brought there, that same era also seeing the establishment of its Mint. Despite the continued existence of the Roman South and East Gates and most of the walls, as well as attributes of an administrative centre, Gloucester seems not to have enjoyed commercial success at this time and boasted no substantial stone buildings. Although not listed as such in contemporary sources, it is likely that the town may have been a Mercian *burh*, functioning as a defended refuge, its defences manned by the *fyrd*, an armed force of levies raised from the local populace on a system determined by land tenure, each Hundred supplying sufficient troops to man an allocated length of rampart. In *c.* 920, Gloucestershire was assessed at 2,400 hides grouped in twenty-four Hundreds. The surviving Roman walls may have been strengthened by earth and timber defences and could have been extended to the river as tenth-century streets cross the line of the western Roman wall which had disappeared by then, while the Roman riverside wall was still standing in the twelfth century. These defences are mentioned in 914 but did not figure in records of earlier refortification, so it has been argued that by the time Gloucester's defences were rebuilt in the years following the Conquest they had already been in existence for possibly 200 years. Offa's priority was to ensure that bridges over the Severn would be held against Welsh incursions. Subsequently, the bridge at Gloucester became the front line in preventing Danish ships sailing up the Severn to attack the Mercian burhs of Worcester and Shrewsbury.

Offa, king of Mercia had established the *burh* of Winchcombe in the second half of the eighth century. In 1007, it was taken over from the Danes by Wessex and became the centre of an administrative area or county. Although this official recognition lasted only a decade, the significance of the town at the centre of its dependent villages lasted until at least the compilation in 1086 of the Domesday Book. The Mercian *burh* was refortified in both the ninth and the tenth centuries and its defences consisted initially of a rampart and ditch. Improvements to this rampart appear to have included having a masonry wall cut into it, possibly in the later tenth or early eleventh centuries when the town assumed its local importance.

Lying on the southern border of Mercia, Bristol grew around the bridge over the Avon which effectively removed Cirencester as a communications node by opening up the Severn valley as the main route to the south-west. Possibly a *burh*, Bristol was the location of a market and a Mint producing coins of Aethelred by 1010. The minster church of St Peter, sometimes thought to have been founded by Offa in the last decades of the eighth century, stood on the future site of the castle. This would have been outside the town's *burghal* defences which are generally held to have occupied the loop formed by the rivers Avon and the old course of the Frome. The major defensive purpose of the *burh* was to defend the bridge over the Avon and prevent the passage of Danish ships up-river. When Alfred reorganised his Wessex *burhs* as a defence against the Danes, he chose Bath, by then, apparently, part of Wessex, rather than Bristol which remained for a while longer in Mercia. Bristol's early street plan nevertheless bears a marked resemblance to that characteristic gridded street layout applied to *burhs* of the time. As Aughton has pointed out, it would have been illogical to site the Saxon town outside of the highly defensible river loop and away from the bridge which constituted its *raison d'être*, gave the town its name, and provided the lowest crossing point of the Avon on the site of, or very near to, the present Bristol Bridge. It seems likely that the earliest fortified Saxon settlement occupied the loop of the Frome and the Avon, with St Peter's Minster located at its neck. Subsequent expansion of the town to the east brought the Minster inside the expanded fortifications, almost certainly constructed of earth and timber. A new bridge over the Frome to the north may also date from this late ninth-century development.

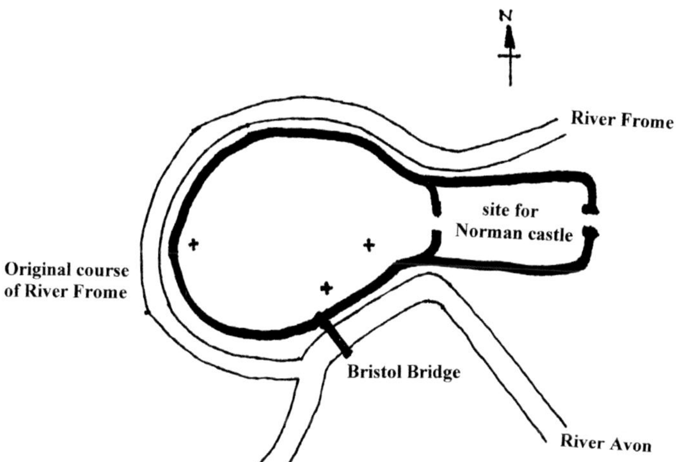

Sketch plan of Saxon Bristol. The Saxon settlement is thought to have occupied the easily defended loop in the river, with a later expansion to the east which would become the site of the Norman castle. Three churches with Saxon origins are shown along with the Saxon bridge.

Mercia extended from Cheshire in the north down to the Thames and the border with Wessex and established defences against their neighbours under kings such as Aethelbald and Offa. Offa's Dyke, along with Wat's Dyke, was built to define the frontier between Powys and Mercia. It runs from the Beachley Peninsula, along the Wye to Hereford and thence along the Welsh border to the river Dee, a distance of almost 150 miles (240 km). The dyke consists of a ditch on the Welsh side with a bank some 8 feet (2.5 metres) high on the English side, the whole being around 65 feet (20 metres) across. There is archaeological evidence to show that some sections were started as early as *c.* 450, and parts of Wat's Dyke in the north, credited to Offa's predecessor Aethelbald who reigned from 716–757, have been similarly dated. It would appear that the discontinuous lengths of Offa's Dyke were dug over a considerable period of time but have all traditionally been attributed to Offa. In Gloucestershire, over forty separate lengths of Offa's Dyke along the Wye valley are listed in the county HER.

Although there may have been some continuity in Cirencester, the Roman buildings were gradually abandoned, and a small village of wooden huts developed near the Roman town. Cheltenham, too, began life as a Saxon village. An Iron Age enclosed settlement at Pucklechurch was reused in Anglo-Saxon times as a hunting lodge. A timber hall is likely to have occupied part of the site and may be represented by the earth platform which bears indications of building works although a trial excavation in the 1960s found no actual evidence. The lodge gained notoriety as the scene of the murder of King Edmund of the West Saxons in AD 946 while apprehending an outlaw. Welshbury Camp may have been refortified after the Romans left if the evidence of piled masonry has been interpreted correctly. While earthwork castles have traditionally been seen as Norman imports, there is very little discernible difference between the fortified residences of Saxon thegns and Norman ringworks. Hillesley, first recorded in 972, has been put forward as such a site although the excavators assigned the site a twelfth-century date.

Deerhurst Church, overlying a Roman building, dates from 804 when the nave, lower part of the west porch, and parts of the porticus were built. Among other features, the upper stages of the west tower date from the tenth century. Within the earlier builds the raising of the west porch to become a three-storied tower is significant in that it contains a doorway, now blocked, asymmetrically placed in its east wall at first-floor level, which accessed a wooden gallery looking down into the nave. Just such a gallery was often provided in churches of this period for the thegn to show himself to his people and was a manifestation of his authority as holder of the symbolic *burhgeat*: the fortified manor and belltower which entitled him to thegnal rank.

Gloucestershire as a Battleground

The *Anglo Saxon Chronicle* records a battle fought near the defended ford at Kempsford in January 800 between the Mercians under Ethelmund and the men of Wessex who were in a hiatus after the death of Beorrhtric, their king, and the assumption of his successor Ecgbryht, who was overseas, visiting the court of Charlemagne at the time. Tradition, not wholly borne out by evidence on the ground or in documents, maintains that Maud's Walk is the remnant of the Saxon defensive work which protected the ford. In 877, a Danish army under Guthrum was loose in Wessex using delaying tactics to avoid having to abide by the arrangements they had agreed with Alfred to maintain the peace and leave the kingdom. They spent the last months of the year in Gloucester before attacking Chippenham, where Alfred narrowly avoided being captured. After spending time regrouping, Alfred raised a strong force and defeated the Danes at the Battle of Edington/Ethandun near Westbury in Wiltshire. Together with the men of Hereford, the men of Gloucester defeated a Danish army in a battle fought in 914. By 959, Mercia and Wessex were again combined under Edgar, and Bristol was the fief of Aethelweard Maew, a descendant of Edward the Elder, Alfred's son. In 1016, Maew's son, Aelfgar, switched sides to support the Danish Cnut in his successful campaign against Edmund Ironside. In 1049, the Welsh king of Deheubarth along with his Danish allies attacked Gwent. Having crossed the Wye at Tidenham, his combined force defeated troops from Gloucester and Hereford. In 1053, he mounted a further attack on Westbury-on-Severn and it has been suggested that a fortification may have been constructed at Newnham-on-Severn as a reaction to such attacks, possibly prior to the Conquest.

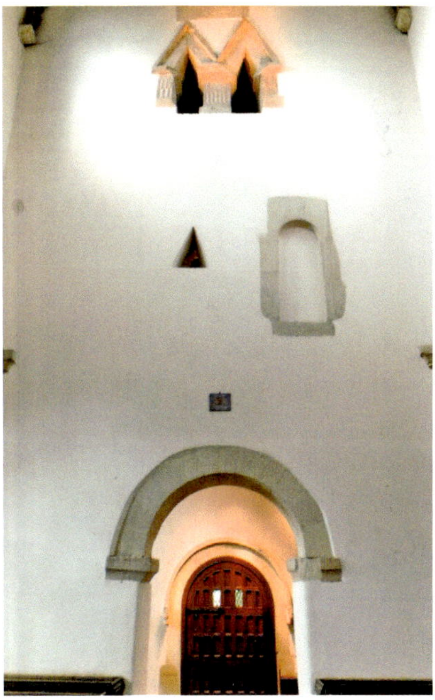

Deerhurst Church, the interior face of the ninth-century tower showing the doorway formerly accessing a timber gallery.

2

Gloucestershire in the Middle Ages 1050–1500

Home to three important royal castles and numerous baronial strongholds, the county provided the backcloth for both momentous events on the national stage and for less significant but equally consequential events for those involved locally. John Goodall's definition of a castle tells us:

> A castle is the residence of a lord made imposing through the architectural trappings of fortification ... [Lord Stafford at Thornbury, for example] ... saw fortifications as central to the character of his residence and his social identity. In a sense it is this understanding of the architectural trappings of fortification, not the quality of the defence they provided, that identify the castle.

There were, nevertheless, times when those very architectural trappings needed to come into play and the defensive rather than cosmetic qualities of a castle might earn their keep.

The Historical Context of Medieval Warfare in Gloucestershire

Bristol submitted to William I after the Conquest and subsequently fought off an attempt in 1067 to recapture the town by two of Harold I's sons, crossing from Ireland to plunder the surrounding area but the town's Saxon defences, and a Norman garrison proved sufficient to repulse the raiders. Gloucester, a key nodal point where roads from every direction converged on the Severn crossing, became a traditional location for the king's Christmas crown-wearing ceremony, probably at Kingsholm Palace, at which, in 1085, William I announced the inauguration of the Domesday Book survey. The earliest Norman castle in the county was Berkeley, described at the time by the diminutive term *castellulum* and built by William FitzOsbern on one of the Saxon estates he had seized. The first castle in Bristol consisted of a motte and bailey with earthwork and timber defences. In 1087, Geoffrey de Montbray, bishop of Coutances, who held Bristol, was part of Robert of Normandy's revolt against his younger brother William II known as Rufus. After the collapse of this rebellion, Rufus imprisoned Robert in Bristol Castle which he then granted to the loyal Robert Fitzhamon who married his daughter to an illegitimate son of Henry I, Robert of Caen, who then became the 1st earl of Gloucester. King Henry died in 1135 having nominated his daughter Maud, wife to the Holy Roman emperor, as his successor. However, Henry's nephew Stephen seized the throne precipitating the civil war, generally known as the Anarchy. When Robert of Gloucester declared for Maud, his half-sister, his supporter Geoffrey Talbot sought refuge in Bristol Castle, and it became their main base for the duration. The castle, allegedly, became the focus of all sorts of brigandage which forced Stephen into action. The stone castle was in process of construction in 1138 with the great keep being built on its mound when Stephen arrived to lay siege to Robert. He quickly found that the castle's location between the rivers Avon and Frome, and its strong masonry defences, made it too tough a nut to be cracked without a greater effort than Stephen was willing to expend. Despite the advice Stephen received to dam the Avon and to blockade the castle with fortified bridgeheads, the siege was abandoned. Miles FitzWalter, constable of Gloucester, switched his allegiance from Stephen and declared for the Empress Maud, escorting her there on her arrival in England in 1139. Miles became earl of Hereford and after his death in a hunting accident in the Forest of Dean in 1143, his son Roger continued his opposition to Stephen. When Stephen was captured in 1141, he was imprisoned in Bristol Castle but was exchanged for Robert of Gloucester who had been taken prisoner by Stephen's troops at Winchester. South Cerney Castle was taken during Stephen's campaign of 1139 when harassment by Miles of Gloucester's forces contributed to his withdrawal. A skirmish, one of many in these unsettled years which often arose from localised feuding, took place around 1140 between the two sides at Beverston Castle. While new strongholds of earth and timber were raised to serve local functions, existing buildings might also be adapted. St Mary's Church at Lower Slaughter provided a base for brigands extorting tolls and protection money from travellers.

After Robert of Gloucester's death in 1147 and Maud's removal to Normandy, the fighting died down as the result of a stalemate. After Stephen's death, Maud's son acceded as Henry II in 1154, granting Gloucester a charter as a reward for its loyalty to his mother. Earl Roger, however, was in rebellion against the king a year later but died that same year, when the town became an important supply base for the Welsh campaigns. The lords of Berkeley who had held their castle through three generations had been dispossessed in 1152 for failing to support the winning side. Henry II then permitted the rebuilding of the castle by the Fitzhardings who established the dynasty which exists to this day. After his accession, part of Henry II's tying up of loose ends involved depriving Robert of Gloucester's heir, William, of Bristol Castle, into which a royal garrison was installed around 1160. William's daughter Isabel married Prince John who was promptly named as heir, a neat way of securing the earldom for the crown, and although the marriage, between first cousins, was annulled, the castle stayed in royal hands, a situation consolidated following the defeat of the Young King's rebellion in 1173. In 1193, John was in rebellion against his father but William Marshal, earl of Pembroke and sheriff of Gloucestershire, ensured that Bristol Castle remained securely garrisoned for the king. In 1171, Henry II had mustered an army at Newnham-on-Severn as a prelude to a campaign in Ireland, possibly basing himself in the castle there.

The dispute between King John and the barons who had invited Louis, the *dauphin* of France, to rule England, saw Bristol Castle garrisoned against them. The Giffards of Brimpsfield Castle sided with the barons against King John who seized their castle and prevailed on the pope to excommunicate them. John died in 1216, in the middle of the Barons' War, and his nine-year-old son was crowned in Gloucester Cathedral as Henry III, and in March 1217, the Giffards had their castle and lands restored. Miners from the Forest of Dean were recruited to serve in Henry III's force besieging rebels holding out in Bedford Castle in 1224, using their specialist skills in undermining the castle's defences, collapsing the walls of the inner bailey and bringing about its capitulation.

In 1233, King Henry III was visiting Gloucester accompanied by some of his most influential foreign courtiers, but Richard Marshal, earl of Pembroke, their chief opponent, whose father had been regent of England and had ensured Henry III's succession, refused to attend the king and was declared a traitor. St Briavels Castle was garrisoned by royal troops as a precaution against capture of the munitions stored therein. Marshal's troops defeated a force led by Baldwin of Guines at Monmouth and a truce was arranged with the king's party. Marshal was then drawn away to Ireland where conflict had been initiated by Peter des Roches, bishop of Winchester and another of Henry's foreign magnates. Marshal was killed in the fighting, and in 1234, some of his supporters in England sought refuge in Flaxley Abbey, supposedly built on the exact spot on which Miles, earl of Hereford, had been killed, possibly murdered, while hunting in 1143. Although a royal signatory to the Provisions of Oxford, laying out the barons' demands for Henry III's power-sharing deal, Richard de Clare, earl of Gloucester, was a discontented man. He had continually been denied the restitution of Bristol Castle by Henry III who bestowed it on his own son, the Lord Edward. By 1259, this dispute had reached boiling point and a party of de Clare's retainers attempted to take the castle by force but were detained by the castle's steward and some of them hanged. De Clare died in 1262 but the next year, the king ordered a complete review of the earldom's holdings before he would allow his son, Gilbert, to inherit. The Lord Edward then found himself besieged by the earl of Derby and the townsfolk of Bristol who were supporters of Simon de Montfort, but tricked his way out, pretending to submit to arbitration, joining his father and leaving a garrison in the castle. By this time, England was fast descending into full-scale civil war. On one side was Henry III with his royal party and on the other were rebellious barons under the leadership of de Montfort and (Red) Gilbert de Clare, the new earl of Gloucester. The king and his barons were in competition to appoint their own as sheriffs and, in 1262, a group of rebel barons drove the royal sheriff, Matthias Bezill, regarded as an alien, out of Gloucester Castle and replaced him. Bezill, in turn, came back and evicted the newcomer. In 1263, John Giffard and Roger de Clifford successfully besieged Gloucester Castle in order to evict this French knight, who was once again sheriff of Gloucester. In 1263, the rebels were successful in capturing Gloucester Castle after a four-day siege during which they had caused widespread damage to the fabric, aided by prisoners, released to bolster the defence (not a procedure to be recommended) but instrumental in admitting the attackers through a postern. One of the rebels, Roger de Clifford was pardoned and made sheriff and constable and along with Giffard, responsible for keeping the peace in the county. The garrison of Gloucester Castle at this time consisted of three mounted sergeants and twenty Welsh foot-soldiers. Soon, de Clifford and Giffard fell out and, summoned to appear before the sheriff of Gloucester, Giffard took a party of armed men with him, gained entrance to the town in disguise, killed some of those present, drove off the sheriff, and, with the help of de Montfort's sons, held the town for their father's cause. The castle, however, held out and was relieved the next spring by the Lord Edward who gained access to the castle from the river, forcing the rebels to withdraw.

The next year, de Montfort, accompanied by the king, whom he had captured at the Battle of Lewes, as his prisoner, arrived in Gloucester hoping to come to an agreement with the earl of Gloucester, formerly one of the foremost leaders of the rebel barons but with whom he had fallen out. De Clare had apparently believed de Montfort to be engaged in an honest attempt to free the country of those foreign appointments which had so incensed his father and the other naturalised barons. He gradually became more and more disillusioned, and the final straw was de Montfort's cancellation of a planned tournament in which he had invested significant sums of money. He appears to have briefed against de Montfort through the Tewkesbury Annals, which were written under the influence of his family, but still attempted to remain uncommitted to either side. Giffard had been captured by the king's side at Lewes, but as the tide turned, he had lost potential ransoms when de Montfort had released captives taken by Giffard's knights, so he, too, chose to change his allegiance once again. De Clare neglected to attend de Montfort at Northampton and, threatened by de Montfort's approach, he and Giffard retreated to the Forest of Dean. They lit a beacon at Flaxley Abbey's grange at Ardland or St White's near Littledean Camp to alert the king's party, held captive in Gloucester by de Montfort's garrison, to their presence. Helped by de Clare, the Lord Edward contrived to escape from captivity in 1265 and proceeded to mobilise royal support around the earl of Gloucester and the Marcher lords to deal with de Montfort. Edward's troops recaptured Gloucester town after a siege lasting fifteen days which involved the use of powerful siege engines to batter the walls. They also took the castle which, safeguarding the last crossing of the Severn available to de Montfort, had been reinforced with a garrison of 300 of his men. Even though the castle there remained in royal hands, de Montfort still enjoyed the support of Bristol's populace which would subsequently cost the town a fine of £1,000. The penalties for this support were compounded when many of the ships loaned by the Bristol merchants to de Montfort for his crossing to Wales were sunk by royal ships sent from Gloucester. De Montfort thus became trapped across the water and was then forced to retrace his steps towards the Mortimer lands on the Welsh border by land. Clifford, having returned to King Henry III's employment, had been ordered to hold the Gloucester bridge against Simon de Montfort's troops and also to destroy the other Severn crossings. However, benefitting from the diversion caused by Edward's attempt to neutralise his back-up force at Kenilworth, de Montfort managed to cross the Severn south of Worcester, but was eventually brought to battle by the Lord Edward and Gloucester together, his army defeated and himself killed at Evesham in the August. Gloucester then chose to change sides again, but his rebellion was brief, and peace was made in mid-1267. Bristol was given up to the king and became the centre of all the local royal estates, with the exchequer housed in the castle. In Gloucester, the garrison was paid off; the burghers were charged with repairing the city walls; supporters were rewarded, and malefactors chastised. In 1276, ships from Bristol intercepted de Montfort's son and daughter as they attempted to land in Wales for Eleanor's planned marriage to the prince Llywelyn. For the moment, the barons had been subdued and in 1278, Edward, by now king, enacted the Statute of Gloucester which challenged the jurisdiction of those magnates who had assumed rights without gaining royal sanction. In 1290, Joanna, second daughter of Edward I, had married Gilbert de Clare, earl of Gloucester. Following his death in 1295, his four children were moved into Bristol Castle to be raised in the setting of an English court, away from the dangers of the Welsh Marches, but still conveniently close to their mother's estates in Glamorgan. The king commanded Bristol's constable to provide appropriate accommodation for his royal grandchildren in houses within the castle walls. Meanwhile, in a secure building and spending his nights locked in a wooden cage bound in iron, built to the king's own specification, was Owain ap Dafydd, the last of the Welsh princely line, who, from the age of seven, would spend upwards of fifty years in close confinement there. Henry, the younger brother of the earl of Lancaster, was pardoned by Edward I in 1307, for aiding the escape of one, John le Harper from a Gloucestershire jail, an incident possibly connected to the conspiracies which underlay much of aristocratic life. A John Le Harper, possibly a son, was in Henry's employ and would accompany his final expedition to Aquitaine in 1345 as an archer.

In 1312, conflict arose between the mayor and burgesses of Bristol and Bartholomew de Badlesmere, constable of the royal castle, over taxes. This culminated in the citizens building a stone wall, cutting off the castle from the town to enforce a blockade. The townsfolk's rebellion stretched over four years but ended in a week-long siege ordered by Edward II. Meanwhile in 1312, Gloucester had once again become embroiled in disputes between King Edward II and his barons when Piers Gaveston, the king's favourite, was executed by the baronial opposition which included the earl of Gloucester. Edward appointed Maurice de Berkeley as keeper of the king's peace and ordered the townsfolk of Gloucester to co-operate with him in putting the city's defences in order. The citizens of Bristol remained in rebellion, and in 1314, the earl of Gloucester was ordered to lead an army against them, but he was diverted to fight in Scotland and was killed in Edward II's ignominious defeat at Bannockburn. The townsfolk held out until 1316 when the sheriffs of Gloucester and Somerset led an army against the town. Maurice de Berkeley blockaded the Avon, depriving the town of its water supply, and siege machines were brought in to smash the wall dividing town and castle, still memorialised as 'Defence Lane'. The town was forced to surrender but discontent simmered for a further six years with rebels being executed in Bristol which continued to be a hotbed of opposition to Edward II. By 1321, Edward was based in Gloucester, in an attempt to pre-empt the warlike preparations of the Marcher lords by crossing the Severn and putting pressure on the Mortimers and their allies who quickly surrendered. In 1322, the Giffards joined the rebellion of Thomas of Lancaster against Edward II and his Despenser favourites and were brought to battle at Boroughbridge and defeated. Giffard and Thomas of Lancaster were among those captured, condemned for treason and sentenced to be executed, Giffard being brought back to Gloucester to be hanged, drawn and quartered. In October 1326, Queen Isabella, effectively estranged from her husband, and most probably already in a relationship with Mortimer, was in Gloucester with her lover. The next year, they led an invasion force into Suffolk. Edward fled westwards to Gloucester, hoping to raise an army in Wales, but with Isabella and Mortimer's army close behind, he sailed for Ireland from Chepstow, only to be driven back by bad weather. The queen and her supporters marched on Bristol where they laid siege to Hugh Despenser the Elder in the castle. Although her troops suffered casualties from Despenser's archers, stationed in pairs on the wall-head, she pressed home successive assaults and the castle capitulated after eight days. Despenser, Edward's long-term favourite and supporter, was promptly hanged, and Isabella rescued her two daughters who had been held hostage. Edward and the younger Despenser were then captured and taken to Monmouth Castle. Hugh Despenser the Younger was executed, and the king was forced to abdicate in favour of his son who was crowned in February 1327. Aware of plans to free Edward, Mortimer handed him over to his son-in-law, Thomas de Berkeley, for safekeeping. After failed rescue attempts and time spent in other places, Edward was returned to Berkeley where he is believed to have been murdered, although there are conflicting and plausible theories about his survival. To all intents and purposes, following embalming and inspection by local worthies from Bristol and Gloucester, Edward's body was buried in an elaborate tomb in St Peter's Abbey, later Gloucester Cathedral, commissioned by his son, an event which led him to confirm the city's charters.

The intermittent warfare of the Hundred Years' War impacted on local families. Maurice IV of Berkeley was wounded and taken prisoner. Although he was ransomed, he never fully recovered and died in 1368 aged only thirty-seven. Other families benefitted from the spoils of the war, enabling them to build such homes as Sudeley and Icomb. While avoiding the devastation caused in France by the free companies which pillaged and destroyed at will, England nevertheless suffered from the attentions of discharged soldiers. The disruption to the social system caused by the Black Death or Great Plague which carried off around half the population had created a vacuum which would easily be filled by rootless soldiery and ambitious barons. Previously prosperous manors were now depopulated, and lords found it easier to plunder neighbours than to rebuild the shattered economy. Subjection to such conditions, and legislation limiting the opportunities for labourers to exploit the labour shortages brought about by the plague, inflamed the discontent which would lead to the Peasants' Revolt. Weak kingship and the corruption of the baronage and the church exacerbated the weakened state of an already dysfunctional society and recurrent outbreaks of plague made things even worse. Strife between the crown and the barons would eventually lead to Richard II being deposed, contributing to further disorder. It is hardly surprising that, in these circumstances, any landowners and others with a stake in society would grasp any opportunity to protect their communities and property by securing them against attack.

While Richard II was in Ireland in 1399, Henry Bolingbroke, supported by many nobles, Thomas of Berkeley amongst them, carried out a coup. Expecting Richard to return through Wales, many of his supporters occupied Bristol Castle. Henry had gone to Berkeley where he joined up with the duke of York who, though regent while Richard was away, had changed sides. Confronted by Henry and York with a superior force, the castle quickly capitulated, and Richard's major supporters were taken and executed, an act referenced by Shakespeare (*Richard II* Act III, Scene I). Soon afterwards, Richard was captured and incarcerated in Pontefract Castle. Thomas Holland, earl of Kent, and son of one of Richard II's half-brothers, and John Montagu, earl of Salisbury, were among those who conspired in the Epiphany Uprising against Bolingbroke. The new king, however, enjoyed widespread popular support and was tipped off, so avoiding capture, and forcing the conspirators to flee to the south-west, but failing in their attempt to seize Cirencester. In Shakespeare's *Richard II* (Act V, Scene VI), Bolingbroke is briefed by his aides: 'the latest news we hear/Is that the rebels have consumed with fire/Our town of Cicester, in Gloucestershire/But whether they be ta'en or slain we hear not'.

Taken and slain indeed they were. Baron Lumley was beheaded in a violent struggle with the townsfolk, while Kent and Salisbury were captured and, after a bungled rescue attempt, summarily beheaded. As Percy soon reported to Bolingroke (Act V, Scene VI): '… the next news is: I have to London sent/The heads of Salisbury, Spencer, Blunt and Kent'.

Sir Bernard Brocas, another rebel captured in Cirencester was taken to Tyburn for execution, and Thomas le Despenser was captured by the mob in Bristol and beheaded on the spot. An alarmed Bolingbroke immediately realised the danger of a live king as a focus for rebellion and, following these preliminary judicial murders, Richard himself was dead within the month.

In the poll tax returns of 1377, Bristol with nearly 6,500 over fourteen-year-olds was only marginally smaller than York and had about a third of the population of London. Even after the ravages of the Great Plagues, it remained England's third city. However, it was Gloucester which retained a greater political significance with Parliament being convened in Gloucester on three occasions. The first, as we saw above, saw the proclamation of the Statute of Gloucester in 1278, whereby Edward I took back those royal powers which his father had been forced to cede to the barons. The second of these was in 1378, when the young Richard II was advised by his uncle, John of Gaunt, that due to the volatility of the mob, the murder of two men in Westminster Abbey had made London an unsafe place. The third time that Gloucester was the setting for the assembly of parliament was in 1407. This was a postponement from the previous year when a French threat had intervened. Welsh Prince Owain Glyn Dwr had been in revolt since 1400, and the prince of Wales, the future Henry V, was dealing with the rebellion with troops levied by the sheriff of Gloucester, and his father wanted to be close at hand. Other events in the area contributed to the necessity of coming to Gloucester at this time: Lady Despenser, widow of the Epiphany Rising plotter, was arrested in Cheltenham while attempting to reinforce the rebels; and Thomas, Lord Berkeley, the English admiral of the fleet, took ships from Gloucester and Bristol to foil a French landing in Pembrokeshire. By 1407, the Welsh had run out of allies and the rebellion was over. The military operations now had to be paid for and the abbots of Winchcombe and Gloucester along with the priors of Deerhurst and Llanthony lent money to the king. Larger sums were loaned by Lord Berkeley and Robert Whittington, sometime shire MP and elder brother of the cat-fancying lord mayor of London. All these and the speaker, son of Geoffrey Chaucer, along with the shire and borough MPs, had supported Henry IV's usurpation of the throne. The Gloucester parliament's stringent enactments ensured that Glyn Dwr lost the support of the people and by 1409, the rebellion had completely fizzled out.

The Battle of Nibley Green, though taking place in March 1469/70 at the height of the Wars of the Roses, had little connection with any events of national importance. It was coincidental that the two protagonists were on opposite sides of the Lancaster–York divide: Thomas Talbot, 2nd Viscount Lisle, was a Lancastrian and William, 2nd Baron Berkeley, was a Yorkist. A long-running dispute over the Berkeley inheritance had seen Berkeley's mother confined to jail in Gloucester where she died in 1452. Her persecutor, the countess of Shrewsbury, died in 1468, leaving her grandson, Lisle, to take up the cause. While Edward IV was busy campaigning in the north, Lisle rashly took the opportunity to issue Berkeley with a challenge to combat. Berkeley accepted and set the morning of the next day for the encounter. This would give Lisle time to scrape together barely 300 supporters, whilst Berkeley could quickly raise upwards of 700 from the castle's garrison, tenant levies, and relatives. Their two private armies joined battle and the result was never really in doubt. As subsequent and much later accounts all describe it, Lisle was forced to risk all in a desperate charge, possibly into an ambush. Berkeley's archers, who included foresters from the Forest of Dean, apparently broke up this charge and Lisle was unhorsed by a lucky (or unlucky) arrow, which entered the open visor of his helmet, and was then swiftly despatched. His troops broke and fled, having suffered up to 150 casualties, and Berkeley's men went on to sack the Lisle manor of Wotton. Subsequent lawsuits, only resolved in 1609, presented Lisle's death as murder and held back on the circumstances of the battle which might, in truth, have been a skirmish involving many fewer participants, and the hot-headed eighteen-year-old Lisle's death might actually have been the legitimised murder alleged in court by his widow.

The Battle of Tewkesbury

On 14 April 1471, at the Battle of Barnet, Edward IV and his brother, George, duke of Clarence, had inflicted a heavy defeat on the Lancastrian army, led by the earl of Warwick who had changed sides and would be killed on the battlefield. That same day, Queen Margaret and the young Prince Edward, heir to his father, Henry VI, landed at Weymouth, aiming to challenge King Edward's grip on the throne. Margaret's forces headed for Wales, intending to link up with their allies led by Jasper Tudor, and despite its Yorkist leanings, stopping off at Bristol to gather reinforcements, guns, and cash, before making a feint towards the strategically placed hillfort at Old Sodbury. Edward IV, meanwhile, was heading westwards from London in pursuit and, falling for the Lancastrian feint, reached Cirencester where he stopped to gather his forces. The Lancastrian army slipped past him and made it as far as Berkeley Castle but still had to cross the Severn at Gloucester if they were to make it into Wales. Sir Richard Beauchamp, the constable of the castle, successfully carried out his orders to deny them entry to Gloucester and, not having sufficient time to take the city by storm, Margaret was forced to bypass Gloucester and head upstream toward Upton-on-Severn via Tewkesbury. Edward had marched his troops through Cheltenham, and at Tewkesbury, he caught up with the Lancastrians who were reluctant to risk a river-crossing while being so closely pursued. Somerset decided to make a stand and a set-piece battle, fought on 4 May 1471, became inevitable. The two armies were evenly matched at around 6,000 apiece, but the Lancastrians enjoyed a defensive position backing on to the rivers. However, the Yorkist army was accompanied by cannon and Edward began the engagement with a prolonged bombardment of Somerset's position. Under such withering fire from both archers and gunners, Somerset's division was goaded into abandoning its secure position and to exploit the congested landscape by attacking through a gap which might have existed between the divisions of Gloucester and Edward IV. However, by exposing its flank to both Gloucester's division and a force of 200 spearmen stationed in Tewkesbury Park, and also unsupported by the rest of the army, it was routed. The whole Lancastrian army then lost its nerve and tried to escape but hampered by the broken country and extensive watercourses, many of those fleeing were cut down. Both Somerset's younger brother, the marquess of Dorset, and the earl of Devon were killed in the fighting along with up to 2,000 of the Lancastrian troops. Prince Edward, the seventeen-year-old heir to the throne, having personally supervised the summary execution of noble prisoners after the Second Battle of St Albans, was himself executed on the battlefield. The two differing levels of Sanctuary were never fully understood or defined, so other prominent Lancastrian leaders were dragged from Tewkesbury Abbey and beheaded with only the briefest of trials. These included Somerset and Devon's younger brother. Some fugitives had made it the 10 miles to Didbrook, claiming sanctuary in the church but they too were executed on the spot. Edward justified the killings by arguing that the Abbey had never, officially, been declared as sanctuary and anyway, that he, being the ascendant king, was perfectly entitled to take the lives of traitors. A subsequent attempt by a Lancastrian force to free King Henry VI from the Tower failed and he died on the very day that Edward's victorious army entered London, his death being laid at the hands of Edward's brother, Richard of Gloucester, later to become Richard III. The battle is commemorated in the local name: Bloody Meadow.

St Briavels Castle functioned as a royal hunting lodge and the headquarters or administrative centre of the constable of the Forest of Dean. It also acted as a repository for munitions produced by the local iron-workings and forges. King John ordered 2,000 crossbow quarrels to supply his mercenary troops who favoured the weapon and in 1223, Henry III procured 6,000 more and sent siege engines constructed in the Forest to the siege of Kenilworth in 1266, along with a siege tower or belfry from his store in Gloucester Castle. Richard I took 50,000 horseshoes on crusade while other products included picks and shovels, and vast quantities of nails. Under Henry III and Edward I, the castle was used as a factory producing tens of thousands of crossbow bolts using local iron and supplying the king and the Marcher lords in their campaigns against the Welsh. Blacksmiths set up a forge in the castle bailey and individuals could produce a daily output of 100 quarrels. In 1408, masons and stonecutters in Harescombe were producing spherical stone shot for the royal cannon.

Medieval Castles

The earliest castles were built of earth and timber and consisted of either a motte or ringwork, each with one or more baileys. The motte was a mound of earth, sometimes based on a natural outcrop, but usually formed from the spoil from its deep surrounding ditch. After the Norman Conquest, mottes were raised in shire towns and surmounted, as they usually were, by a timber tower, they were intended to overawe the townsfolk. Ringworks were little different from the thegnal households they succeeded, consisting of an enclosure containing a hall and other domestic buildings surrounded by a bank and ditch with a palisade and a defended

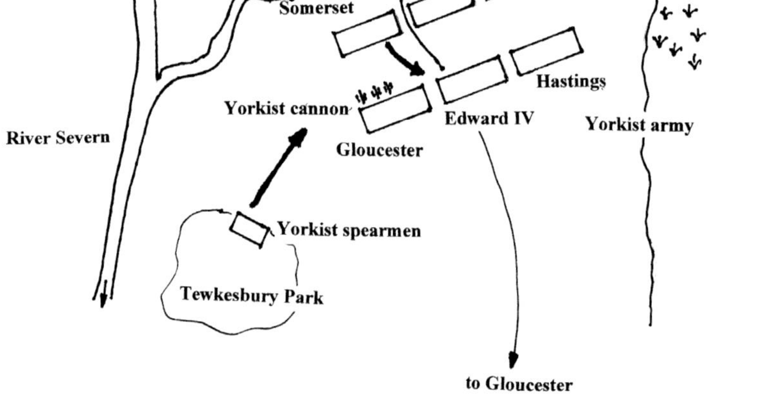

The Battle of Tewkesbury (after Seymour) showing how Somerset's rash attack on the Yorkist centre led to disaster for the Lancastrian army.

gateway. Both forms would normally have an attached court or bailey containing other buildings such as stables, forges, and workshops. Gloucester and Bristol Castles both started as urban motte and bailey earthworks established by trusted followers of William I in the period between 1066 and 1086. Other such castles included the original site at Brimpsfield, later to be replaced by a stone castle on a different site.

As we have seen, Gloucestershire was a battleground during the period that is usually referred to as the Anarchy of Stephen's reign which lasted from 1135–54. While Winchcombe's *burghal* defences survived as the basis of a castle in 1141, a large number of earthwork castles were built from scratch and fought over during this time, particularly during Stephen's campaigns against the earl of Gloucester in 1139–42. These included motte and bailey castles at Dursley, said to have been built by Roger of Berkeley after his loss of Berkeley Castle in 1153 with Drakestone Point being suggested as a possible site; South Cerney, Taynton (Old Church), Tetbury, and the apparently unfinished motte at Bledisloe. At Dymock, also sometimes referred to as Pauntley or Castle Tump, Aylesmore, a motte and bailey castle was granted to William de Braose *c.* 1148–54. The motte stands 45 feet (14 metres) high and has a summit diameter of 25 feet (8 metres), and the bailey, defined by a bank and ditch, stands 8 feet (2.5 metres) above the surrounding area. The Tump may have had an additional function as a Hundred meeting-place as had Bledisloe. A site at nearby Castle Farm shows traces of a moat and bridge. At Hailes Wood, 500 metres east of the abbey, there was a ringwork which occupied an Iron Age enclosure, reused in the twelfth century as a castle. Here, a roughly D-shaped ringwork surrounded by a bank up to 7 feet (2.3 metres) in height and 5 feet (1.5 metres) in depth with a counterscarp bank in places, possibly formed part of the Hailes Castle, in existence between 1138–50, but whose exact location has never been fully established. Another ringwork at Newnham-on-Severn enclosed a level area 113 by 65 feet (35 by 20 metres), originally surrounded by a bank and ditch. A resistivity survey reported in *Glevensis* 48 (2015) suggested the presence there of a well and a stone building. Other ringworks were built at Littledean, where the rampart contained significant amounts of stone possibly representing the foundations of a curtain wall, and a raised platform, once described as a 'guard point', was noted: Soudley (East Dean) and Stowe and Taynton (Huntley) where the circular earthwork measures 100 feet (30 metres) across. Taynton (Old Church) began life as a ringwork but was soon converted into a motte with two baileys. A supposed ringwork at Hillesley was excavated in 1979 to reveal a horseshoe-shaped earthwork of twelfth-century date, whose rampart had been revetted in timber on the outside and with drystone walling internally. These defences were strengthened over the next two centuries and the outer revetment was replaced by a 4-foot-thick (1.2-metre-thick) wall of un-mortared masonry around the outer edge of the 23-foot-wide (7-metre-wide) rampart. Shortly after this, the site was abandoned and given over to other uses when stone buildings were added.

Most probably some or all of those earth and timber castles of indeterminate age at Upper Slaughter, where excavations found pottery from the eleventh to the thirteenth centuries, Haresfield, Lasborough, Hewelsfield, and Tewkesbury (Mythe), date from the twelfth century. Fortifications at Cirencester were built by the empress next to the church in 1141 and subsequently burnt and the earthworks levelled by Stephen with excavations in 1964 finding that the west end of the abbey site was overlain by a twelfth-century courtyard. Very little remains of any of these castles, and many were around for only a very short time, quickly falling into disuse once Henry II got a grip on the country. Some of these may have had masonry elements as at English Bicknor whose motte was found to contain a chamber measuring nearly 12 feet (3.6 metres) square, and where an outer bailey took in the church and may have constituted a village enclosure. The castle continued in occupation into the thirteenth century, with a specific mention in 1217. One of two apparent earthwork castles at Newington Bagpath, has been interpreted as a collapsed Norman tower on a motte constructed of small stones quarried from the surrounding ditch. Old Ley Court at Westbury, not to be confused with Westbury Court, is another possible motte and bailey castle. Dixton Castle is recorded as occupying an Iron Age hillfort, but both monuments have had doubt cast on them. A castle at Tetbury, sometimes identified as occupying the Iron Age camp near the church, is mentioned as being attacked by Stephen in 1144, but it has also been pointed out that Beverston is often referred to as Tetbury. The existence of a castle at Harescombe may stem from misreporting by William of Worcester in 1480. A stone manor-house may have stood there but a confusion appears to have arisen with neighbouring Haresfield.

A very few earthwork castles were developed into masonry fortresses. Bristol, Gloucester, and St Briavels, all acquired by the crown by 1200, were important royal castles from the twelfth to the fifteenth centuries. Gloucester and Bristol were both built at the lowest bridging point on their respective rivers: the Avon and the Severn. As key bridging points could not easily be bypassed, resident garrisons could sally out to intercept either invading armies or rebel forces, making it difficult for them to link up. The River Severn, especially, represented a significant impediment to mobility, consistently limiting both attacks by the Welsh and home-grown threats to the throne. This can be seen in both the operations against Simon de Montfort culminating in his defeat at the Battle of Evesham in 1265, and in the role of Gloucester's garrison in preventing the Lancastrian army from crossing the Severn leading to its defeat at Tewkesbury in 1471. Medieval kings spent a great deal of time on the road and, along with only half a dozen other favoured defensible locations, both Bristol and Gloucester contained residential royal apartments. All three castles were shrieval castles, important providers of the infrastructure for sheriffs' administration of justice and Bristol was also one of the five castles chosen by King John as dispersed treasury locations in his attempt to maximise tax collection and it would have housed a staff of clerks to maintain the records. Castles also played an important role in the logistics of medieval warfare. St Briavels was primarily a munitions factory and store; siege engines were stored at Gloucester; and Bristol was used to house stockpiles of smaller weapons. Bristol and Gloucester match the stereotype of the royal castle in that each had Great Towers probably built in the reign of Henry I in the early decades of the twelfth century. These were keeps on the scale of Dover or Norwich with great halls on one or more floors, audience chambers and other public space rather than domestic accommodation. Henry II is thought to have built the smaller square keep at St Briavels which would have had rooms of a more domestic nature and fewer grand rooms. Henry III spent money on both Bristol and Gloucester a century later. As well as these royal keeps, there were similar keeps at baronial castles. Brimpsfield's keep would have been similar in style to Gloucester's but half the size, while Lydney's was even smaller, nearer to that at St Briavels. A different style of keep can be seen at Berkeley where the motte forms the core of a shell keep wrapped around it. The motte at Miserden also carries traces of a shell wall built around the lip of the motte and replacing an earlier timber palisade. Here, the motte measures 130 by 110 feet (40 by 34 metres) and 30 feet (10 metres) high, with a rock-cut ditch 30 feet (10 metres) wide and 20 feet (6 metres) deep. The bailey is defined by a bank and ditch, and a gatehouse was excavated at the base of the motte. Even at fifteenth-century Sudeley, the Dungeon Tower performs, if only visually, the function of the keep, dominating the rest of the castle and its surroundings, making a powerful statement about the owner's status. Royal and baronial castles functioned as garrisons, homes, law courts, depots, arsenals and armouries as well as, at royal castles, places where the king might exhibit himself, putting on appropriate shows for the benefit of his subjects. Although Gloucester and Bristol underwent six sieges between them during these centuries, medieval warfare generally sought to avoid large-scale actions whose outcomes were often unpredictable. Kings and barons alike preferred to raid their enemies' lands; to harass their enemies' tenants, neighbours and allies; and to destroy their crops, livestock, and dwellings. This ravaging might be financed by ransoms and protection money and castles provided secure bases from which to pursue these activities.

Bristol was chartered in 1155 and it became an important trading centre for the import of wine, particularly from France, and for the export of dyed and woven cloth. Access to the port was improved in 1239–47 by diverting the River Frome, creating a new channel 750 yards (700 metres) long. The first castle in Bristol, begun possibly as early as 1068, consisted of a motte and bailey with earthwork and timber defences. Its situation between the two rivers has been cited as an example of a castle whose strength was derived more from its location than from its structural defences. The stone castle was in process of construction in 1138 with the great keep occupying one quadrant of the partly demolished motte and probably being completed by 1141. It was around 80 feet (24 metres) square, on three floors divided by a cross-wall supporting two roofs, side-by-side. There were square turrets at the corners, one of which was higher than the others, and a forebuilding. It may have been surrounded by a separate wall, parts of which containing a gate-arch have been uncovered but only fragments of part of the base of the keep now remain. Along with half a dozen other similar sites, Bristol's ditches were deepened and widened on the orders of King John. Henry III added a barbican to the western gate, a great hall and a new gate-tower. The vaulted porch still visible, whose dimensions fit exactly with a contemporary description, is thought to be that of the great hall, built to a similar size as that surviving at Winchester: 110 by 55 feet (34 by 17 metres). In the southern curtain fronting the river there was a gatehouse with twin D-shaped towers and a short length of riverside curtain wall with two arrow-loops and a postern, unusually accessed via a tunnel, can still be seen. The castle's East Gate has been excavated and revealed to have been a simple square gate-tower of *c*. 1100, with side walls and a timber sill. Following the castle's occupation by Simon de Montfort's troops in 1265, repairs were carried out. New chambers were later built to a high specification with glazing and collapsed lengths of curtain wall were replaced over the centuries, but by 1480 the whole site was sliding into ruin although Leland, around 1535, describes towers 'yet standynge'. Anything remaining after the Civil War was ordered by Parliament to be slighted in 1655.

Gloucester stood at one of the most important points in the country's road system, and shortly after the Norman Conquest, its castle was built. This stood in the south-west corner of the Roman walls, as a motte and bailey, displacing sixteen houses in the process. The motte stood in the north-west corner of a rectangular bailey estimated as measuring 360 by 260 feet (111 by 80 metres) and their conjoined ditches followed the line of the Roman rampart and ditch on two sides. The motte, with a base diameter of 165 feet (50 metres), was built on sand and despite attempts to stabilise it with large amounts of clay it would always remain unstable and would be abandoned after barely fifty years' occupation. Any buildings on either the motte-top or in the bailey would have been of timber. Indications of a possible hall-like structure have been found in the centre of the bailey now under Ladybellegate Street. When the castle site was shifted onto firmer ground to the west, the motte was lowered and reformed as a barbican and by 1239 the original bailey had been taken into the precincts of the Dominican friary. The new castle was begun around 1120 with a large, free-standing, rectangular hall-keep measuring around 95 by 65 feet (30 by 20 metres), on three floors with a higher turret at each corner, and a possible fore-building. One corner of the keep, which was demolished in 1785 to rebuild the prison, has been uncovered during building operations and excavations in 2015. A further eight houses were destroyed to clear the site for the keep, inner and outer baileys and other outworks. One gatehouse, with a drawbridge over the

Bristol Castle, the porch of the vanished Great Hall.

Bristol Castle, the unusual postern which emerges from underground.

Bristol Castle, arrow-loops in the curtain wall alongside the riverbank.

moat, connected to the town on the north-east curtain of the castle and another accessed a bridge over the Severn, defended by a timber brattice on the far bank. A third gatehouse, again bridging the moat, was built in 1257 to open onto the road to Llanthony Priory. Henry III spent regular sums of money on repairs and improvements, so, along with the keep, a number of other domestic buildings occupied the surrounding courtyard, including the great hall with its associated kitchen; apartments and chapels for the king and queen and, after 1251, for their son, the Lord Edward, the future Edward I. Towers were added to the walls, repaired or replaced as necessary, and the ditches cleaned out. Following damage caused during the assaults on the castle of 1263–65, money was spent on repairs to the Severn bridge and on the walls. The keep had been subject to the most severe damage being partly gutted by fire which burnt through two floors, and although considerable sums of money were spent between 1270 and 1438, the keep never really recovered and was held to be in danger of collapse for much of that period. In 1791, it was finally demolished to allow a new prison to be built on its site.

St Briavel's Castle appears to have been conceived as an earthwork castle, the successor to an earlier ringwork at Stowe, and was owned by Miles, earl of Hereford, a supporter of the Empress Maud during the Anarchy but was taken into royal ownership in the 1160s and it was during the reign of Henry II that the keep was built. This is said to have stood 100 feet (30 metres) high, but something nearer to 70 feet (20 metres) might be a more realistic height. It measured 51 by 45 feet (15.6 by 13.9 metres) and was a solar tower on a roughly square plan with a vaulted basement and rooms on two or three floors. King John added a domestic block with a hall and royal apartments within a buttressed curtain wall, to which Henry III added a chapel. The imposing twin-towered gatehouse was rebuilt in 1292 for Edward I, but its spurs point to a local influence emanating from South Wales, echoing those at neighbouring castles such as Goodrich and Chepstow, which became a feature of other Clare fortresses. Another departure from the Edwardian norm is the gate passage defended by an unusual sequence of three portcullises, with further such barriers in the entrance passage's side doors. The basements, spurs rising from semi-octagonal plinths and the presence of portcullises with semi-circular grooves are features shared only by St Briavels and the Clare castle of Tonbridge in Kent, as are the polygonal internal tower fronts at all levels, suggesting a shared designer and mason. The gatehouse at St Briavels falls into the category of a keep/gatehouse on the chamber block model providing commodious accommodation at first- and second-floor levels, a feature, again shared with both Edwardian and Clare castles, particularly in Wales. Unlike Tonbridge, however, where the spiral stairs mirror themselves, those at St Briavels, all rotate clockwise. Also renewed at the time of Edward I, were parts of the curtain wall, surrounded by a wet moat, now a garden, fed by a spring.

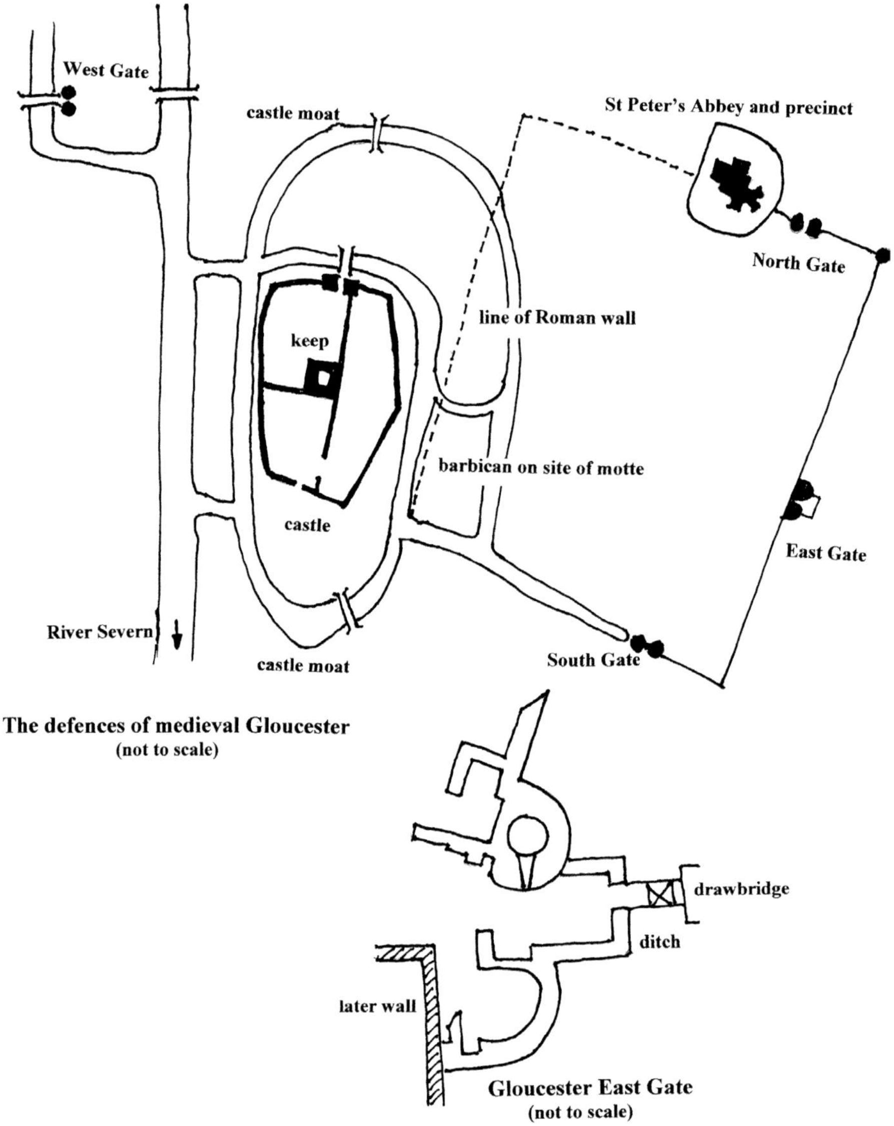

Sketch plan of the defences of medieval Gloucester with a detail of the East Gate.

Like Bristol and Gloucester, Berkeley Castle also began life as a motte and bailey but after 1154 the motte was encased in a stone wall up to the height of the motte. Three semi-circular turrets, one of them containing a vaulted chapel, were then added on the southern and eastern sides and wall and turrets together were built up to a height of 60 feet (18.5 metres). The keep was entered via a forebuilding which was basically a two-storey covered passage following the curve of the shell-wall with a doorway at each end, that on the inside being 20 feet (6 metres) above the courtyard level. A curtain wall enclosing the Inner Bailey with a stone hall built against its south-eastern side followed from 1160. Following the dramatic events of the early-1300s, the castle was largely rebuilt from 1342 by Thomas (the Rich) of Berkeley who stayed onside with the young Edward III. He built both the outer and inner gatehouses, the latter formerly with a barbican with octagonal turrets, a new hall/kitchen/chapel range and on the line of the shell-wall the bulky fourteenth-century Thorpe Tower. This replaced a fourth turret and may have been built as much to provide palatial accommodation at the heart of the castle as to counter the threatening proximity of the roof of the church. In fact, it was from that very platform that the castle was bombarded in 1645, although the detached church tower, rebuilt on its medieval plinth, had been positioned a good way away from the body of the church to minimise the danger of the castle being overlooked. Details

St Briavel's Castle, the gatehouse.

in the domestic parts of the castle reflect the care taken to achieve perfection. John Goodall suggests that Thomas employed the same mason for the castle as for the family's foundation of St Augustine's Abbey, now Bristol Cathedral. Details such as the polygonal arches used in the great hall are clearly influenced by the abbey where they can be seen in the Berkeley Chapel. Writing in the *Castle Studies Group Journal* (No. 32, 2018-9), Neil Guy also put forward an argument for a distinct Severn Valley design of arches, citing Berkeley's inner gatehouse and others of the castle's features, Beverston and St Briavels as examples, all dated to around the first third of the Fourteenth Century. The medieval gatehouse adjacent to the current visitors' vehicular access to the castle was removed from Yate and re-erected here during the early twentieth century.

Some baronial castles were also developed from their earth and timber beginnings. The first earthwork castle at Brimpsfield (SO946127), near Ermine Street, was built around 1086 but had been abandoned by the mid-twelfth century. The oval, flat topped motte measures 115 by 65 feet (35 by 20 metres) but is only 8 feet (2.5 metres) in height. A stony bank around the perimeter of the motte top may indicate the presence of masonry defences and a causeway crossed the ditch at the south-west corner. A new castle was built some 500–600 yards to the west (SO941128) by the Giffards, sometime around 1140. It was a ringwork measuring 525 feet (160 metres) east to west and 360 feet (110 metres) north to south, subdivided by a cross-wall with a gatehouse giving access to the walled inner ward and surrounded on two sides by a wide dry moat. There was a large central tower or keep, around 60 feet (18.5 metres) square with a forebuilding, and a gatehouse tower at the south-west corner of the ringwork with further towers at the other three angles. Following the Giffard brothers' rebellion against King John in 1215, the castle was sequestered but returned to them by Henry III. John Giffard's participation in the rebellion of Thomas of Lancaster, defeated at Boroughbridge in March 1322, resulted in his execution. His castle was slighted at the express command of Edward II and almost all traces of it were quickly obliterated. Only earthworks and a tiny fragment of the gatehouse with part of a portcullis groove now remain on site.

Lydney Castle, on a strong hilltop site and now reduced to its foundations, is generally regarded as a solar keep, that is a smaller square measuring 48 by 56 feet (15 by 17 metres) rather than the larger rectangular hall keeps such as Gloucester or Bristol. The keep, which retains the base of what may have been a stair-turret at its north-west corner, sat across the north-east corner of a roughly square enclosure with a turret at its opposing angle. The gateway in the eastern wall is sandwiched between the keep and a substantial rectangular tower, approached through an outer bailey, defined by earth banks, lying to the east. The castle is thought to date from the time of the Anarchy but is clearly a much more robust structure than the majority of its contemporaries. It may have had some connection to the local charcoal and iron-making activities of the Forest.

Above: Berkeley Castle, the inner gate and shell wall.

Right: Berkeley Castle, the doorway into the keep.

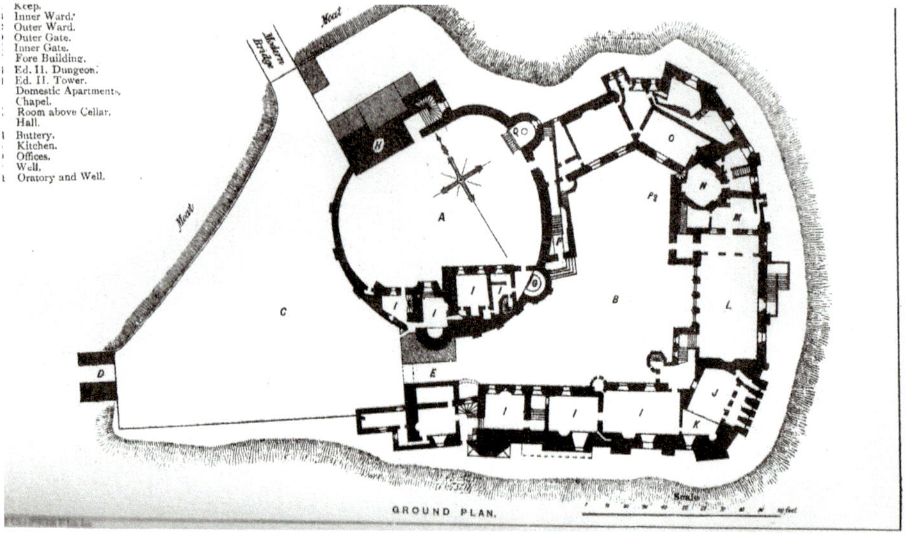

Above: Berkeley Castle, ground plan from GT Clark, *Mediaeval Military Architecture Volume 1* (London: Wyman & Sons, 1884).

Left: Berkeley Castle, an example of the castle's distinctive internal archways.

Maurice de Gaunt (or Gent) was granted a licence to crenellate a castle at Beverston in 1229. His original castle may have been trapezoidal with round towers at the angles and possibly, a simple gateway. Sometime in the period 1330–1361, the castle was partly rebuilt by Thomas de Berkeley, around 1348–50. The two circular towers on the west side which had been linked by a first-floor hall were demolished. That on the north was replaced by a diagonally set, three storey square tower, further altered in the next century, and that on the south by a linking block to a new south-west solar tower. The existence of earlier buildings provided the opportunity to create a multi-level domestic range. The solar tower contains a vaulted chapel on the first floor, with a chamber and tiny oratory above, all accessed by a spiral stair in a turret. It is connected by several sets of stairs to a range of chambers, joining up with the reorganised north-west Tower, completing this western block. At the eastern end of this complex is a roughly triangular bailey at whose apex is Berkeley's gatehouse with twin drum towers and a chamber over the gate-passage with its portcullis. The foundations of a further round tower have been found in the rectory garden on the castle's former north curtain, suggesting the original castle had a pentagonal trace and the whole site was originally moated. A later house now occupies the southern range which was probably the site of an earlier hall. Although there is a tradition that Lord Berkeley financed his improvements out of the spoils of the French Wars, it is more likely that his fortune came from wool. It is said that on one occasion, 5,000 sheep were shorn in the castle precincts.

Ruardean Castle originated in the late eleventh or early twelfth century and may have consisted of little more than earthworks with timber buildings. Alexander of Bicknor, a royal cleric appointed treasurer of Ireland in 1307, received a licence to crenellate in 1311. It would appear from investigations that he built his new castle, which consisted of two ranges of buildings on opposite sides of a walled quadrangular enclosure, on top of a flattened motte, revealed by the LiDAR survey. A twin-towered gatehouse stood in the south-east corner and a circular tower at the south-west. This tower, 26 feet (8 metres) in diameter, described as a keep, retains an octagonal vaulted basement and is the only recognisable element remaining. An outer enclosure may represent the bailey of the earlier castle.

The site of Sudeley's earthwork castle, which changed hands at least four times during the Anarchy, is unknown so it may have been where the later castle would be built, or it might have been nearer to Winchcombe. Ralph Boteler apparently made his fortune from the wars in France and spent it on a new quadrangular castle around 1450. The southern or Inner Court measured 160 by 130 feet (50 by 40 metres) and had a tower at each corner, with a hall and other domestic buildings ranged around the courtyard. The Dungeon Tower at the south-west corner was regarded as the keep, already an outdated concept but still regarded as a necessary symbol of lordship. Within twenty years, Boteler had forfeited the castle which was given by Edward IV to his brother, Richard of Gloucester,

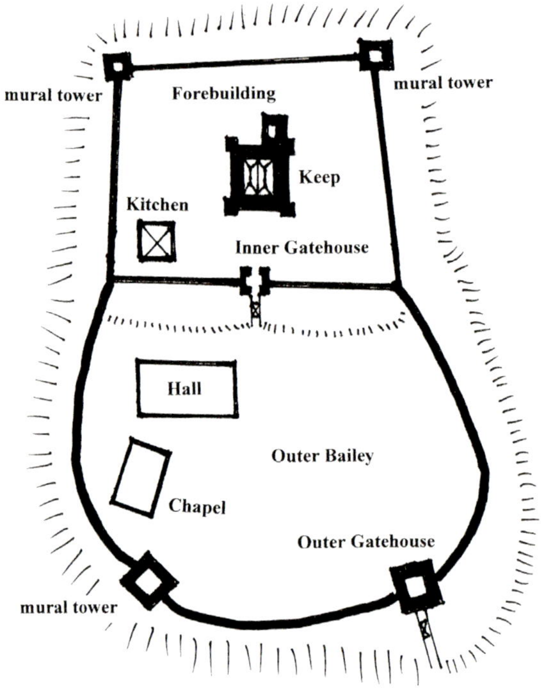

Above: Brimpsfield Castle, a conjectural reconstructed ground plan (after Dodd & Moss in *Glevensis* 25).

Left: Beverston Castle, the principal tower and stair turret.

later Richard III, who was responsible for the surviving great hall and linked apartments befitting its royal owner. It would appear that Boteler had added the northern or Outer Court at a slightly different angle to the earlier one, connecting the two with a simple gateway in the cross-wall which linked the Garderobe and Portmare Towers. This Outer Courtyard was completely rebuilt by Lord Chandos in 1572, keeping only the shell of the gatehouse in the north wall. Following damage during the Civil War, partly due to the parliamentarian garrison using the panelling for firewood, and subsequent slighting in 1649, the castle remained a ruin until restoration was begun in the first half of the nineteenth century. The surviving window tracery in the former great hall is evidence of the high quality of the craftsmanship of Richard of Gloucester's masons.

Henry I was beset by rebellious Norman barons, one of whom, his brother-in-law Hugh of Montfort was captured in 1124 and imprisoned in Gloucester Castle 'placed in vicious bonds' and kept in custody for fourteen years. By 1185, Gloucester Castle was in general use as a prison and in 1196 money was spent on new fetters. Castles were often used as jails for those awaiting trial, and as prisons, especially for royal or aristocratic prisoners who could be kept confined but generally in the luxury to which they were accustomed or entitled. Eleanor of Brittany, sister of Prince Arthur, supplanted and probably murdered by her uncle, King John, was kept for long periods of time at, among other places, Bristol and Gloucester, from *c.* 1202 until her death in Bristol Castle in 1241. She spent several years in Gloucester Castle where a new fireplace was installed in her apartment in 1238. She was also allowed visitors, both noblewomen and priests, and was well supplied with food including venison for herself, her two maids and her two household knights. If an important prisoner escaped, the constable and sheriff, often the same man, were held responsible and punished. Prisoners captured by Edward I in his Scottish expeditions were brought to Gloucester and St Briavels to be held there until their ransoms might be paid. Edward II was imprisoned in Berkeley Castle for six months prior to his supposed murder in September 1327. St Briavels, on the other hand, was the seat of the warden of the Forest of Dean and served as a jail for those caught poaching or otherwise falling foul of the stringent forest laws prior to their trials. At the end of the medieval period, the gatehouse then became a debtors' prison, remaining in use while most of the other buildings had, by the end of the eighteenth century, collapsed. An underground, circular chamber beneath the gatehouse, discovered in the 1880s, was identified as a dungeon or pit prison. This has been dismissed as wishful thinking and is now reckoned to have been a secure storeroom or strongroom.

Strong Houses and Fortified Manor Houses

As well as those castles which were either built solely for military purposes or were expected to be capable of fulfilling military roles, there were, throughout

Sudeley Castle, the fifteenth-century dungeon tower, part of the castle's first build.

England, many medieval houses which exhibited elements of defensibility. They would often be moated; provided with a gatehouse set in a surrounding, usually embattled, wall; furnished with a strong tower; and be built in stone over vaults. Controversy has raged for several decades over the significance of the royal grant of a 'licence to crenellate', fourteen of which were granted in Gloucestershire between 1153 at Berkeley and 1510 at Thornbury. Recipients ranged from Robert FitzHarding who was being rewarded for his adherence to the party of Henry II; to clerics, public servants, barons, and knights; through to the duke of Buckingham. Receipt of a licence was often perceived as a token of royal approbation which might confer bragging rights among peers or neighbours. Crenellation could mean anything from a few battlements to ostentatious towers with machicolation and arrow-slits or gunports. Some of these places such as Beverston, licensed in 1229 to Maurice de Gaunt, would probably qualify as a fortress, while Thornbury, licensed in 1510 to Edward Stafford, 3rd duke of Buckingham, is generally thought of as a palace whose martial embellishments were intended to be recognised as signals of power, status, and heritage, largely symbolic and affording little value as a defensive work. Many of those strong houses such as Wick Court at Arlingham, Withington, or Wanswell were subsequently transformed into polite Tudor or Georgian manor houses but may retain vestiges of their earlier identities.

Holm(e) Castle, Tewkesbury, was a manor of the earls of Gloucester from the twelfth to the fourteenth century and stood on a site previously occupied in the late Saxon period. An early Norman hall was replaced in the thirteenth century by a grand, stone first-floor hall over a vaulted undercroft. The hall stood in a courtyard with other buildings constructed of timber on stone footings, some of them built against the stone wall of the enclosure which was defended by timber brattices and a gatehouse. A possible tower, around 10 feet (3.5 metres) square, standing beside the hall, may have contained a tiny chapel, accessed at first-floor level from the hall. The manor was frequented by King John who had married Isabel, the heiress to the earldom of Gloucester.

At Thornbury, part of the earlier fortified house has been excavated next to Stafford's later castle in what would become the Privy Garden. It is a possibility that a licence was granted in the fourteenth century but has not yet come to light. The house consisted of a great hall and chapel on the eastern side of a walled courtyard which was entered through a gatehouse on the west and enclosed all the usual domestic offices. Barns, granaries, and stables occupied an outer enclosure. Henry Stafford, 2nd duke of Buckingham, was executed in 1483 and the house was forfeit but returned to the family in 1498, becoming the 3rd duke's principal seat. His redevelopment, modelled on Henry VII's Richmond Palace, retained the hall and chapel as the eastern range of a new inner court. The new western range was planned to consist of a symmetrical range of four towers, the central pair flanking the gateway. These towers and two intermediate turrets were all polygonal, following what had become the fashion in southern England, but only the octagonal South-west (or Lord's) Tower was ever completed to full height. An outer court on the west contained two ranges of lodgings for the duke's household retainers. While the western façade of the inner court with its sturdy towers projects a formidable image, the southern front is composed of apartments with large, full height oriel windows looking out onto gardens, originally furnished with timber cloisters. Although Stafford received a licence to crenellate in 1510, permitting him to create a 'fortalice or castle', one of only fifteen licences to be issued after 1485, all along he had been suspected of harbouring unacceptable aspirations. The lodgings were seen as the barracks of a private army, something which the insecure Henry VII had discouraged with legislation in 1504, and the defensive features included machicolation, arrow-loops and crude gunports. The conclusion to which the king was drawn was that this palace was clearly intended to simulate, if not constitute, a fortress. It was left to Henry VIII to have Stafford executed in 1521 leaving his great building project unfinished. Thornbury is often cited as the last medieval castle to be built, perceptibly drawing on the accepted vocabulary of fortification.

On a much smaller scale than Thornbury or Sudeley but still wanting to achieve the look were several other Gloucestershire manors. The originally moated and embattled Icomb Place dates from *c.* 1230 but the present building of *c.* 1420 has a buttressed projecting embattled gateway leading into the first of two courtyards, separated by a great hall with late medieval and early Tudor features. One end of the hall abuts the solar range, while the other accesses the kitchen and buttery through the screens passage. The south range has been demolished. It was built by Sir John Blacket who fought at Agincourt and, like Boteler at Sudeley, may have financed his grand house from ransoms or loot, and was killed in France in 1430. It is currently (2021) undergoing extensive restoration work as a family home. Similar in both scale and appearance to Icomb was Kempsford Castle, built on a Saxon earthwork in the twelfth century on the bank of the River Thames, near to where the church presently stands, to guard the ford. It came to Henry Plantagenet, earl of Lancaster, in 1298, through his marriage to Maud de Chaworth and became a retreat for them from the turmoil of Edward II's reign. Their son, Henry of Grosmont, 1st duke of Lancaster, bestowed it on his religious foundation in Leicester. His heiress, Blanche of Kempsford, married to John of Gaunt in 1359, was mother to Henry IV, and patroness of Chaucer. The castle survived the Dissolution but was demolished around 1610 to be replaced by a house which was itself demolished *c.* 1790. Most of the building stone was removed but the present manor farmhouse may stand on the original castle's cellars. It was probably no more than a moated manor house with two courts, one entered by a gatehouse and containing a hall and detached kitchen. Hasfield Court began life as a moated medieval house with a gatehouse whose archway survives, built into the stable block. Withington manor, licensed in 1271 to Godfrey, bishop of Worcester, who had been lord chancellor from 1266–68, had a gatehouse and was mentioned in 1476. Cromhall, licensed to William de Wauton in 1311, may have been a weak tower house, raised by a soldier with connections to the de Berkeley family but whose main estate was in Essex. Olveston Court was a house of the Denys family, dating back to the thirteenth century, which preserves important features of a developed moated manor. There was a main courtyard and a subsidiary kitchen courtyard, partly surrounded, in the fifteenth century by an embattled curtain wall incorporating a wall-walk as well as windows and doorways and an early Tudor gatehouse. The defensive works are confined, by the lie of the land, to the south and west. It was owned from 1475 by John Walshe, an administrator of royal and baronial interests. Standish Court was built in the fourteenth century for the abbot of Gloucester where a fourteenth-century gatehouse with two arches, buttresses, and part of a spiral stair survive. The standing structure known as King John's Castle at the

Right: Thornbury Castle, the Lord's Tower, at the south end of the uncompleted façade.

Below: Icomb Place, the fifteenth-century east front with its castellated gate tower.

Mythe in Tewkesbury represents the stair-tower of a medieval house which may have been a lodging of the abbots. It is a three-storey (formerly four), square tower with thick rubble walls, abutting a Tudor house, which has had many additions right up until 1991. There are blocked medieval doorways at each level confirming its function as a stair-tower and there was also a free-standing chapel. Yate Court was a fortified house from the time of Edward I. Licensed in 1299 to John Wylington, it stands in an oval moat, 80 by 100 yards (75 by 90 metres). Visiting antiquarians in 1898 noted a vaulted gatehouse with inner and outer arches, portcullis groove and a first-floor guardroom, confirmed by Edwardian-era photographs showing a stone gatehouse. The courtyard was surrounded by the ruins of ranges of buildings and a round and a square tower. Ruins of the gatehouse and a hall remain within a moat, but most of the house had been razed in 1644 after occupation by parliamentarian forces. Wylington and his brother were both taken prisoner serving Edward II at Bannockburn in 1314. They then joined Edmund of Lancaster's revolt and, like Giffard, were captured at Boroughbridge. John was able to buy a pardon, but Henry was executed. The gatehouse now stands at Berkeley next to the modern visitors' entrance. A house known as Culverden, probably Willington Court in Sandhurst, licensed in 1318, was also owned by the Wylington family whose Henry received the licence. The later house, which retains the remains of a medieval stone tower, may have originally been moated. Robert de Sapy, warden of the Forest of Dean and constable of St Briavels Castle, was licensed to enclose his manor of Old Ley Court in Westbury on Severn, with a fence, and to build a timber peel-house in 1330, parts of whose moat are still visible. De Sapy was a household knight of Edward II and had held administrative posts in Durham and elsewhere. The licence may have been granted by Edward III as a gesture of reconciliation. Knole Park, in Almondsbury, a manor owned by the Fitzharding family, was a house of *c.* 1570 built by a mayor of Bristol, overlooking the Bristol Channel. Incorporated into a modern house is a fifteenth-century octagonal tower, with an oak spiral stair, which was part of an earlier structure owned by the abbey of St Augustine, now Bristol Cathedral. The mayor's house has now been redeveloped but the tower lives on.

Medieval Moated Sites and Homestead Moats

The late Iron Age/Romano-British farmstead enclosure, excavated at Miserden in 2014, was found to be surrounded by a deep ditch and, most likely, a palisade. This would have been more for the protection of livestock than a full-on defence against hostile neighbours. This approach to security must have characterised many of the later sites of a similar nature which can be found in the county dating from the medieval period. Gloucestershire is rich in medieval moated sites

Standish, the internal arch of the gatehouse of the manor house built by the bishops of Gloucester.

Tewkesbury, The Mythe; doorways at three, formerly four, levels identify this much modified structure, attached to a Tudor house, as the stair tower of a medieval building, probably built for the abbot but known as King John's Castle.

or homestead moats, many of which can be found in the Vale where abundant supplies of clay were available for lining wet moats. Aberg includes seventy-nine moated sites in his county list with seven of them scheduled as ancient monuments. In the 1978 list of excavated moated sites, Gloucestershire is represented by just three: Prestbury, Leckhampton, and King's Stanley. A preliminary checklist published in 1978 by the Gloucester & District Archaeological Research Group, published in *Glevensis* 12, lists 114 but with twenty-five tagged as only 'possibles'. The HER records sixty-two medieval moated sites in the Forest of Dean alone, but one-third of these may be considered spurious or at least doubtful. The Dean Archaeological Group survey listed eighty-six water features within the Forest of Dean District Council area, of which some forty-odd appear to be authentic moated sites while the rest are fishponds, garden features or drainage ditches.

Moated sites come in all different shapes and sizes and were dug for a wide range of uses. They were owned by abbots, prelates, holders of public office, knights, and yeomen. Moats could be dry but most, given their secondary purpose of supplying fishponds, or draining water away from habitation, found their watertight properties significant. The main purpose of a moat was to surround a platform on which a house might be built, as at Wightfield Manor, Deerhurst, the original house of Sir John Cassey with its moat dating from the fourteenth century. Podsmead, a grange of Llanthony Priory in Gloucester, was a square moated site containing a house used by the prior with the usual hall, parlour, chapel, kitchen, and a detached gatehouse. Newark House nearby was built in the fourteenth century by the prior of Llanthony to rival the abbot of Gloucester's house at Over, the Vineyard, a three-sided moat with a slope on its fourth side. Wanswell Court Farm, at Hamfallow, is an unusually large moat surrounding the house which dates from 1256 when a chapel was licensed. The nucleus of the present house is a hall-house of *c*. 1450–60, standing on an island measuring 340 by 180 feet (105 by 55 metres) while the maximum width of the moat, which was created by damming a stream, is 240 feet (74 metres). Leckhampton's moated site represents one of the three manors in the village. The fourteenth-century Leckhampton Court, held by the Giffards from 1322, consisting of a hall range and two other ranges around a courtyard, was the grandest, and this site may be its precursor. It could also have been the centre of the Broadwell manor. It consists of a platform surrounded by a pentagonal ditch. An excavation in 1933, having noted earlier observations of masonry footings each side of the ditch which had dry-stone walls as revetments, uncovered a square framework of four well-preserved oak beams which spanned the moat and clearly represented a bridge. On the island, stone walling was found indicating the presence of buildings and this was backed up by building debris including roof tiles and dressed masonry. Pottery dated the bridge to no earlier than the 1300s, but older pottery fragments indicated earlier occupation on the site. A resistivity survey carried out in 2004 suggested the presence of a gatehouse and possibly a two-storey hall/solar block. Evidence has been found for the moat's continued occupation over successive centuries after Leckhampton Court had come into use. Prestbury was a house of the bishops of Hereford and occupied the westernmost of a pair of conjoined moated sites. The site is known from excavation to have been used from at least the twelfth century and possibly for three centuries before then. Pottery of the eleventh century has been found and the bishops had owned the property since before the Conquest. The house, which stood on the western platform consisted of a stone-built ground-floor aisled hall with a timber upper storey. At the upper level were the solar and chapel. In the western part, along with gardens, lay the detached kitchen, linked to the main block by a covered way. The eastern platform contained stables, barns and fishponds. The nucleus of the house and the kitchen dated from the twelfth or thirteenth century, and later additions included a porch/gatehouse which enclosed a small, inner cobbled courtyard. Both enclosures together were surrounded by banks, and ditches which would flood in the winter. A moated site at Lower Noverton, Prestbury, may represent a semi-fortified house known as 'The Court' and owned by the prior of Llanthony, but the existence of the fourteenth-century hall house known as 'The Priory' near the church would tend to make this unlikely. This may, therefore, be an earlier house of the Baghots who built Hall Place. A hall, 65 feet (20 metres) long and kitchen were uncovered in the 1960s. Forthampton Court retains the great hall of *c.* 1380, built within the moat, for the abbots of Tewkesbury. Next to the churchyard of Moreton Valence lie two rectangular moated enclosures measuring 145 by 100 and 145 by 60 feet (45 by 31 and 45 by 18 metres). Royal permission to fell ten oaks was granted by Henry III in 1253 for the construction of the hall of a house which may have occupied the larger island, but which was derelict by 1372. Sir John Cass(e)y, who died in 1400, had been chief baron of the exchequer in the reign of Richard II. He inhabited Wightfield Manor in Deerhurst, a stone house on a moated site, rebuilt in the sixteenth century. Cranmore Farm, Beverston, has produced pottery from the twelfth and thirteenth centuries along with stone roofing tiles suggesting that a building occupied the island. Not all buildings within moats were so substantial. The moat at Whitcliff Deer Park, Ham and Stone, encloses a small island, 90 feet (27 metres) square, which may have been either a hunting lodge or a keeper's dwelling. Some, so-called empty moats, may have been used to preserve cash crops such as fruit or honey, from pilferage. Everes Farm, a trapezoidal homestead moat, 280 by 170 feet (86 by 52 metres), at Redmarley D'Abitot, shows no evidence of buildings at all.

Moats were predominantly rectangular, but six circular moats were reported in the county in the list published in *Glevensis* 11, including Scar Cottage at Redmarley D'Abitot, Mote Farm, Moorend Green, and The Mount, at Haresfield churchyard, described as a circular or almost square moat around an island measuring 160 by 155 feet (50 by 48 metres), which may have been the home of the sheriff of Gloucester *c.* 1100. Matson moat was oval. At Coleford, the moat known as Kinnersley Court in 1334, then as Brecknocks/Breckness Court, was investigated by Bristol University, and three oval, rectangular, and semi-circular platforms were uncovered. The Knight's Green homestead moat, measuring 90 by 125 feet (28 by 38 metres), and Aubreys Place, Bromsberrow, a moat in good condition around an island 80 feet (24 metres) square, mentioned in 1424, are two examples of several in the county of trapezoidal shape.

Moats ranged in size from Barons Court at Churcham measuring only 60 by 25 feet (18 by 75 metres) to one at Blockley, Aston Magna Village, measuring roughly 325 feet (100 metres) square. Barrs Court at Oldland is a large, moated site which retains most of its water-filled ditch. The interior platform measures 390 by 340 feet (120 by 105 metres), surrounded by a moat up to 50 feet (15 metres) wide and 4 feet (1.2 metres) deep with an outer bank. Within the enclosure stood numerous structures, many built in masonry, which in 1540 still included a lodge and a chapel. Access was across a causeway and over a drawbridge. There were fishponds, a deer park, a warren and a dovecote. Some parts of the county saw quite dense concentrations of moats. There were nine moated sites in Taynton alone, two of which have been filled in whilst another survives only as two arms, and a fourth as entrenchments. Of the survivors, one was circular, one rectangular, and another polygonal.

Ecclesiastical Buildings with Defensible Features

In his introduction to the Gloucestershire volumes of the *Buildings of England* series, David Verey, the local authority on Gloucestershire's buildings, highlights a group of Norman churches which lack east windows: Notgrove, Aston Blank, Baunton, Winstone, and Brimpsfield. He suggests that they were continuing a Celtic tradition which saw churches as vulnerable to attack and therefore sought to limit the possible access points. At other Norman churches in the Cotswolds, east windows were of restricted size. The practice may have evolved from early places of worship which found the nave added onto the western face of the priest's windowless dwelling. Verey also cites two examples of defensible church towers. One is at Kempley where St Mary's is described as having a West Tower of *c.* 1276 'built for defence with no entrance from the outside'. The second is the church of St Peter, St Paul, and St Mary at Westbury-on-Severn where he describes the detached, three-storey tower of *c.* 1270 as 'a garrison or watchtower, heavily buttressed at the angles', and standing 50 feet (15 metres) away from the very slightly later church. The tower, with its octagonal stair-turret, stands on a battered plinth. A spire was added in the fourteenth century and a medieval chapel, demolished in 1862, adjoined the tower on its east side. St Nicholas at Ozleworth has a rare hexagonal central tower, dating from the 1130s with what Verey describes as 'blocked round-headed loopholes' in two faces.

As we have seen, prelates, abbots, and senior clergy, like their temporal peers, also had appearances to think of and took reasonable precautions to secure their homes against unwelcome visitors: witness the fourteenth-century gatehouse at Standish Court, a house of the bishops of Gloucester. Apart from those defensible residences we have seen above, the majority of monastic foundations were also given walled precincts with limited access through gatehouses. This may have been more a way of controlling access and egress rather than anything more overtly defensive, but security would, nevertheless, remain an issue throughout the medieval period. St Peter's Abbey, later Gloucester Cathedral, was built to the north of the Roman walls in 1104–13 by Abbot Peter who walled its precinct, but the later, existing wall dates from 1218. A length of original northern precinct wall along the south side of Pitt Street, now forming the boundary of King's School, was part of the Abbot's Lodging or Bishop's Palace which was demolished in 1860 and had been regarded as integral to the city walls until the mid-fifteenth century. The thirteenth-century Great (or St Mary's) Gate was the main outer gate to the abbey, with the fifteenth-century Inner Gate close by. Other gates include the fifteenth-century St Michael's (or Cemetery) Gate, of which the front portion of the lower storey, and the later upper stage in brick remain, and the fourteenth-century King Edward's

Westbury-on-Severn, the detached, solidly defensible tower of the church of St Peter, St Paul & St Mary, dated to *c.* 1270.

Gate, with a turret and walling with blind panelling. Other precincts existed at the Gloucester Greyfriars and Blackfriars. The city of Bristol was also well-endowed with walled monastic establishments: the Augustinian St Augustine's Abbey, now Bristol Cathedral, the Priory, Greyfriars and White Friars, plus the Temple Church at Temple Meads. The restored abbey's Great Gate, dating from *c*. 1165 still stands with its sixteenth-century superstructure. The gate of 1180 at Cirencester with its coach and pedestrian arches; the fourteenth-century gatehouse of Kingswood Abbey with ranges to each side; and the fifteenth-century Abbot's Gate at Tewkesbury, all survive. In 1373, a licence to crenellate was granted to the abbot of Winchcombe and this may have been for a gatehouse and precinct wall which have long since disappeared. Part of the medieval precinct wall of St James's Priory was uncovered during excavations on the Bristol Royal Infirmary site. The gatehouse of Quenington Court, a preceptory of the Hospitallers retains a thirteenth-century postern incorporated into the later gatehouse. At Llanthony Priory, Gloucester, there are the remains of the outer gatehouse of 1500 as well as remains of the precinct wall and inner gatehouse incorporated into a later farmhouse. At Westbury-on-Trym, the mid-fifteenth-century college of priests, part of a grandiose scheme for a proposed cathedral, was built as a quasi-fortified quadrangular structure with round corner turrets and a massive castellated gatehouse. This became a manor house after the Reformation and the present building retains the gatehouse-tower and two of the corner turrets. A gateway with the arms of Katherine of Aragon remains in the western court of Prinknash Abbey.

Town Walls

Medieval towns were generally walled as a way, not only of providing an element of security against attack, but also of preserving trading rights by controlling access and ensuring that residence qualifications were observed. Towns were permitted to raise revenue to maintain their walls via murage grants from the crown, which sanctioned the levying of local taxes. It was the responsibility of the town's burgesses to maintain the walls, and at Gloucester, in 1266–7, they were ordered by the king, to expand part of the town ditch, necessitating the demolition of several properties. Records of a series of murage grants for Gloucester exist from 1226 up until the 1360s, and for Bristol, from 1232 until the mid-1460s. Some town defences were very rudimentary, and Winchcombe's burghal defences survived into the medieval period, while Tetbury is thought to have enjoyed the protection of earthen defences of post-Conquest date. Stow-on-the-Wold may have had free-standing gates, serving more as tollbars, to assert the town's commercial status.

Gloucester's Roman walls formed the basis of their medieval successors only on the eastern and southern sides, but the city's northern expansion was marked by

Above: Gloucester, the remaining wall of the Bishops' Palace forms a stretch of the original precinct wall which dates from 1218.

Right: Gloucester, the internal arch of the thirteenth-century great gatehouse in the abbey precinct wall.

Above left: Bristol, the gatehouse of the Augustinian abbey, whose lower stages date from *c.* 1165, is crowned by upper floors of the sixteenth century.

Above right: Kingswood, the fourteenth-century abbey gatehouse is flanked by monastic buildings.

Below left: Tewkesbury, the fifteenth-century Abbot's Gate into the abbey precinct.

Below right: Gloucester, the outer gatehouse of Llanthony Priory.

Right: Westbury-on-Trym, the massive gatehouse of the mid-fifteenth-century college of priests built with ostentatiously defensible features.

Below: Cirencester, the abbey gatehouse, dating from 1180, has coach and pedestrian archways.

free-standing gates without connecting walls and the Cathedral precinct, overlying the north-west corner of the Roman walls, provided protection at that point, with the river offering further protection in the north-west. The Normans used the walls of the Roman gate passage as the basis for their East Gate. By the thirteenth century, twin asymmetrical towers had been added to the face of the wall, linked by an angled barbican. The northern gate tower was more drum-shaped and projected further into the field than the southern tower. A chamber above the gate was used as a meeting-room by the municipal authorities and in the 1500s, the northern tower had a chamber built onto the old Roman gate passage wall behind the drum tower, which became the Bridewell. The North Gate was used as a jail with an adjoining lodging for the jailer. As we will see, the walls formed the basis for the city's defences in the Civil War and, after the Restoration, Gloucester's defiance of Charles I was reflected in a royal order to demolish the city walls.

Bristol had never been a Roman town, so the city's walls originate only from Anglo-Saxon times and, in effect, were laid out from scratch in the thirteenth century, an excavated bastion in King Street being dated to 1250–1300. Situated between the rivers Avon and Frome, Bristol presents a strong but irregular urban layout, tied to the needs of an important port, dominated by mercantile interests. Beginning in the 1240s, an ambitious engineering programme involving the re-routing of the Frome, to facilitate the reclamation of marshy ground, dictated the line of the walls beyond their original, central nucleus. The walled area of the city was greatly expanded within the new Marsh Wall, linking the two channels of the Avon. Another of Bristol's several extensions was defended by the Portwall which pivoted on Tower Harratz on the banks of the Avon. This large, circular tower of unique design included arrow-loops which could command both the broad ditch and the riverbank. Underground passages connected the tower to sluice gates controlling the water level in the ditch and may have accessed the Watergate and a sally port. The Portwall took in the suburb of Redcliffe and exhibits an advanced defence system which incorporated over twenty clusters of closely linked arrow loops. This level of provision may be explained by the ownership of the area by the militant Order of Templars and the monumental gate which allowed access through the Portwall to the Templars' extramural holdings provides further emphasis of their martial preoccupations. A less imposing gate in the northern walls accessed the slipways and quays lining the Frome. Bristol's medieval churches built over town gates provide an echo of arrangements in other similar towns such as Warwick but the church of St John, one of four churches located by or over gates, remains the only one to survive. Whilst its present form, with a central archway flanked by pedestrian passages, represents a Victorian reconstruction, the fan vaulting and portcullis groove in the central section are original. The Newgate, like gates in other medieval towns, served as a jail. Such expansions of the earlier walled areas to include new suburbs was quite unusual for medieval towns, but Bristol was unique in being chartered as a county in 1373 under Henry III, independent of both Gloucestershire and Somerset, the culmination of royal grants proceeding from

1172. However, absorbing these new areas by the construction of successive lines of defence, would prove, in the Civil War, to have produced an unmanageable enceinte. Ironically, in 1658, following the slighting of the castle, a new gate was built to guard the town's eastern approaches, and the Temple and Redcliffe gates in the Portwall were rebuilt in 1730, complete with porters to keep an eye on traffic.

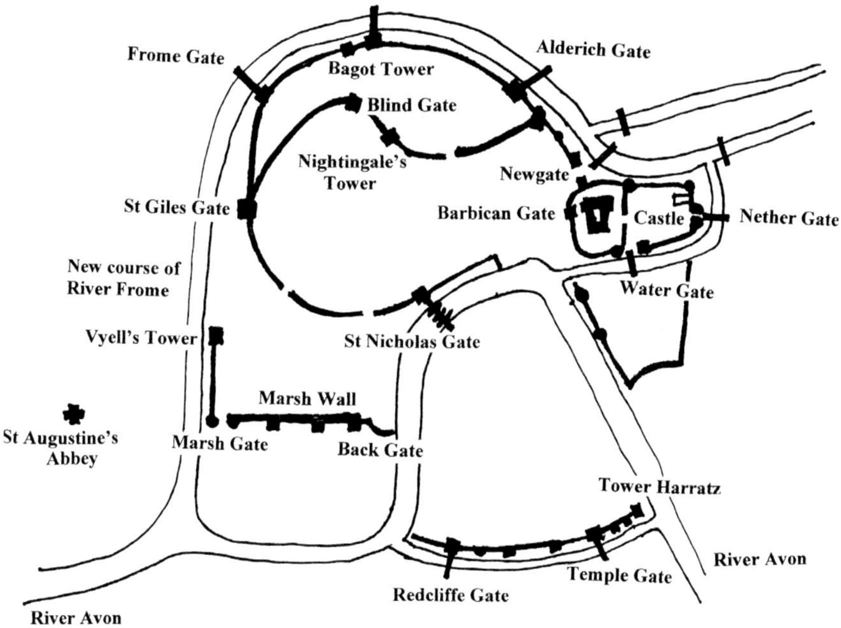

Simplified sketch plan of the defences of medieval Bristol (after Fry).

3

Gloucestershire in Tudor, Stuart, and Georgian Times 1500–1840

Across southern England, a system of fire beacons, established in Anglo-Saxon times, was maintained on inter-visible sites to relay warnings of invasion. These used church towers or other elevated sites but in hilly country hilltop sites were preferred, with intervals of around 6 miles (10 km) being deemed practicable but given their dependence on weather conditions for their effectiveness, beacons were only permanently manned between March and October. These long-established chains were used right up until the time of the Armada threat in the 1580s. Placenames can offer clues and those derived around *twt* or *toot* signifying mounds may be significant here. The Tutt in Hewelsfield, lying east of Offa's Dyke, is cited as such a possible example, located to warn of incursions over the Welsh border. Any site with a wide viewshed could serve as a beacon but, clearly, the highest hilltop sites would offer the widest views. The highest spot in the Malverns was the Worcestershire Beacon at 1,394 feet (425 metres) and Cleeve Hill, at 1,083 feet (330 metres) is the highest point in the Cotswolds. Haresfield Beacon stands at 700 feet (215 metres) and Painswick Beacon at 920 feet (283 metres). Other known beacon sites include Chedworth Beacon, Wyck Beacon, Primrose Hill in Lydney, and Robinswood Hill, south of Gloucester. The lighting of the beacons would prompt the lord lieutenant to order the militia to assemble and for the weapons stored in the county's armouries to be distributed. When Queen Elizabeth I visited Bristol in 1574, she was treated to a pageant featuring a four-day-long mock battle. However, martial entertainments aside, Bristol was predominantly a mercantile centre which was able to supply only four suitable ships for Drake's fleet, which sailed against the Spanish Armada in 1588.

Certain features of castle architecture continued to be applied, even well into the seventeenth century and as late as *c.* 1630, Stanway House, built for the Tracy family, was provided with an imposing gatehouse. However, those few medieval fortresses which survived provided venues for predominantly peaceful events. The discovery of a banqueting house in the garden at Sudeley Castle where Elizabeth I was entertained in 1592 was reported in 2019.

In 1539, Henry VIII ordered a muster to assess the number of armed men each shire might be able to raise. This assessment was carried out Hundred by Hundred with Gloucestershire's thirty odd Hundreds producing totals ranging from 886 able men in Barkeley to sixty-two in Tybolston. Able men were then classified as archers or bill-men, and the availability of weapons and armour was also explored and recorded, with an inventory being compiled. In 1553, Queen Mary I repealed the Statute of Winchester whose requirement to maintain what was essentially medieval weaponry had become obsolete. In 1588, in view of the threat from the Spanish Armada, Elizabeth I was granted leave to impress men for the militia. Gloucestershire had the potential to produce 14,000 able men of whom only some 3,000 had received training, and actually supplied ('furnished') 4,000 men of whom 1,000 remained untrained and 300 were pioneers. Additionally, 235 cavalry were mobilised consisting of twenty armoured lances, 180 light horsemen, and thirty-five petronelles who were armed with pistols braced against the riders' chests (*poitrine*).

The Civil War in Gloucestershire

While many neighbouring counties were controlled by royalists throughout the wars, the citizens of Bristol and Gloucester, though sandwiched between the king's capital of Oxford and royalist South Wales, nevertheless declared for parliament. In those disputed counties of the south Midlands, towns and villages suffered from the constant plundering, levies, taxation, and outright banditry of marauding armies. It made little difference where an individual's sympathies might lie if the troops approaching were seeking provisions, livestock, plate, or cash, and even shoes, bedding, and clothing, and resistance was an invitation to murder, theft, arson, and the total destruction of livelihood. Citizens in towns fared no better and if a siege was threatened, they would often suffer more from their 'protectors' than from their attackers. Houses outside the walls would be destroyed to provide clear fields of fire and those inside to facilitate the movement of troops or the siting of cannon. Large amounts of money and goods would be levied from the populace to pay the troops who would be billeted in the town's houses at the townsfolk's expense. It has been estimated that up to a third of Gloucester's buildings were destroyed and around 20–30 per cent of Bristol's population perished as a result of their three sieges. Thus, collateral damage characterised the new concept of total war which no one could avoid. The dominant strategy throughout the war was the possession of territory, and in attempts to ensure that they maintained their grip on newly won territory, both sides established garrisons, often in small towns or isolated country houses. Campden House had been built in 1613 outside Chipping Campden. It was established as a garrison in 1645 by the royalist Colonel Bard. He issued dire warnings that if the local villagers failed to deliver their assessed contributions by a specified day, they might 'expect a troop of horse [who will] fire your houses without mercy, hang up your bodies wherever they find them, and scare your ghosts'.

Further, they should expect no protection from the parliamentarian troops in nearby Tewkesbury. When Bard left the area, he burnt the house down to the ground to prevent its occupation by the enemy leaving only the flanking pavilions which remain today. A number of other prominent structures were severely damaged or destroyed during the course of the war. These included Kempsford Bridge, rebuilt in 1647 after its destruction in the fighting, and Taynton Church. This had been destroyed and a new one, oriented north-south, was built by order of parliament in 1647–48. Place names may commemorate historic events during the wars. Bloody Acre Camp at Cromhall is supposedly named after a massacre in the Civil War. Cromwell House in Naunton, had been built in *c.* 1600, and later named for his leader by a later occupant. It belonged to Richard Aylworth who is credited with impeding the progress of Charles I at Stow-in-the-Wold in 1644, when he was attempting to manoeuvre his army around the Evesham area.

Chipping Campden, one of the two pavilions of Campden House which remained after the main house was destroyed in the Civil War to deny parliamentarians its use as a garrison.

Local dignitaries raised regiments for whichever side they favoured. Sir Ralph Dutton of Standish, deputy lord lieutenant of Gloucestershire and younger brother of Sir John, the county's MP and heir to the Sherborne fortune, raised a regiment in July 1642. Initially recruiting from his own county, he appointed local men: John Palmer, possibly of Dorington, as senior captain and Thomas Browne as ensign. Picking up recruits across the Midlands, he led his regiment to join the king in Nottingham. He took part in the Battle of Edgehill and was present at the capture of Cirencester and the sieges of Bristol and Gloucester. In December 1643, he relinquished his colonelcy to Sir Stephen Hawkins, a professional soldier who had fought in the continental wars and retired to the royal court at Oxford. In 1646, his Standish estate having been sequestered, he left England to live abroad but was drowned when his ship foundered. Henry Stephens raised a Gloucester Regiment in April 1643, possibly from the militiamen of the Trained Bands and the company of volunteers raised in the previous July. Stephens was captured in June 1643 near Stow-on-the-Wold, so his Town Regiment came under Governor Massey's command.

Bristol in the Civil War

The city of Bristol remained one of the most important cities in the country occupying a strategic position at the mouth of the Severn, controlling access between the West Midlands, South Wales, and the Southwest, and trading with Ireland, Spain, and France. The city was occupied by a parliamentarian force in December 1642 but following the royalist victory at Roundway Down in July 1643, it became vulnerable to attack. A royalist force of 14,000, led by Prince Rupert who had established his HQ in the college in Westbury-upon-Trym, elected to storm the city with its 4-mile-long (6.5-km-long) circuit of defences defended by an entirely inadequate garrison. This included only 1,800 soldiers; the Bristol Trained Bands in three companies, 100 strong, with a total of 210 musketeers and ninety pikemen; the Artillery Company formed of volunteers in 1625; and up to 1,000 armed, but untrained, townsfolk under their governor, Nathaniel Fiennes. Although the attacking royalists sustained heavy casualties, sheer weight of numbers as they attacked in force from two directions, and the defenders' consequent inability to concentrate their men at multiple points of attack, meant that surrender quickly became inevitable. De Gomme's account of the assault quotes Fiennes as saying that recognising that his men were becoming disheartened and were running low on powder and shot, he thought about setting fire to the town and retiring into the castle, but decided, or was persuaded, against it. Following fighting which lasted barely four days, the surrender shocked the country with the suddenness of the collapse of an apparently well-defended city with a competent governor, Fiennes being later threatened with execution for his perceived failures. The medieval city walls already enclosed a large area including the suburb in the loop of the River Avon defended by the Port Wall, pivoted on the Tower Harratz. Here, a long, rectangular redoubt with a V-shaped end, furnished with musket-loops and embrasures for cannon, has been uncovered, further strengthening the medieval defences at this vulnerable point. However, once the suburb of Redcliff had been taken in and earthwork forts and linking walls superimposed, the resultant circuit demanded enormous numbers of defending troops. Forts were built, as Richard Israel has demonstrated, utilising the high ground surrounding the city on the west and north, on Brandon Hill and Windmill Hill with other redoubts at further points right around the city. These strongpoints were often quite formidable, with cannon mounted on bastions; ramparts as much as 30 feet (9 metres) above the bottom of the ditch; and storm poles and *abbatis* to hinder the assaulting troops. The city's weakness lay in the ramparts and ditches linking these strongpoints. Not only were there insufficient troops to man them adequately but the banks were barely of sufficient height to shield a defender and the outer ditch no more than 5 feet (1.75 metres) deep. This allowed the attackers to pierce the defences between the forts on the western side of the city, by using grenades and fire-pikes. An assault on the southern side of the city was repulsed but once the royalists had penetrated the western defences, they were free to roll up the defenders from inside the city. Israel's analysis of the landscape has produced viewsheds which show the limitations of the cannon, ostensibly positioned to prevent infiltration but deficient in range.

During the subsequent royalist occupation, Bristol was the port of entry for several thousand Irish troops arriving from Cork. The opportunity was also taken to strengthen the earlier fortifications, under the direction of Bernard de Gomme, the foremost royalist engineer. New bastions, with stone revetments were constructed at Brandon Hill Fort and a completely new Royal Fort replaced the old Windmill Hill fort. This was pentagonal with five angle bastions and a grand gateway all built in stone. A half-moon bastion may have been constructed within the existing ditch to protect Lawford's Gate. In 1644, over 100 cannon were mounted on the defences and these were increased to 140 the next year. However, even though Prince Rupert had had two years to effect his improvements, there remained weak spots. Hedges and other cover outside the defences had not been efficiently cleared to create effective fields of fire. Some defences had been effectively strengthened but others were 'much decayed'. Nevertheless, Rupert had laid in sufficient stores to feel confident of being able, barring mutiny (his word), to sustain a siege of up to four months' duration.

Oppoite above: Sketch plan of the defences of Bristol in the Civil War. Concentric lines of medieval walls protected the nucleus of the city and its close suburbs whilst a girdle of forts, redoubts and gates with defended outworks was thrown around the whole city creating a perimeter too long for effective defence.

Oppoite below: Bristol, the Royal Fort of which only vestiges remain.

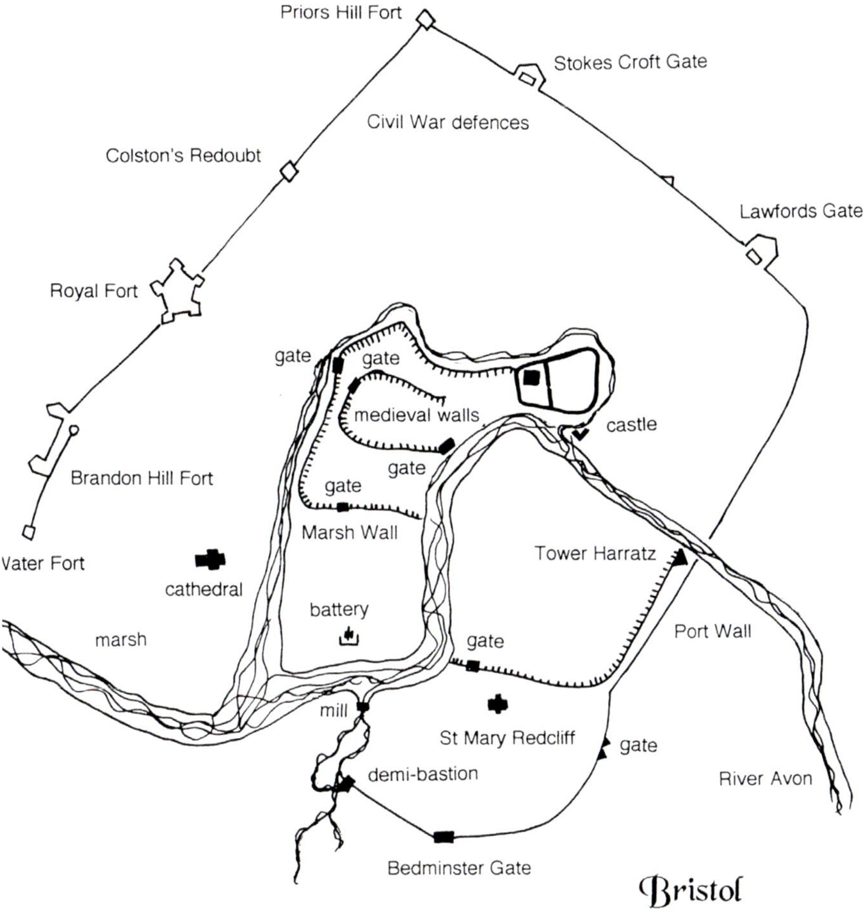
Priors Hill Fort
Stokes Croft Gate
Civil War defences
Colston's Redoubt
Lawfords Gate
Royal Fort
gate
gate
medieval walls
castle
Brandon Hill Fort
gate
gate
Marsh Wall
Vater Fort
cathedral
Tower Harratz
battery
marsh
gate
Port Wall
mill
St Mary Redcliff
gate
demi-bastion
River Avon
Bedminster Gate
Bristol

Counter to these positives, the waning royalist cause had left Bristol isolated from any hope of relief by a field army and furthermore suffering from plague and low morale. Rupert's garrison of 1,500 soldiers and two regiments of Trained Bands, perhaps 2,500 men altogether, would prove even less capable of defending the circuit of defences than had Fiennes' troops in the previous siege.

In August 1645, Fairfax, accompanied by Cromwell, approached the city with the bulk of the New Model Army. Summoned to surrender the city, Rupert played for time by affecting to negotiate. Fairfax was having none of this and began a bombardment of one of the outlying forts at 2 a.m. on 10 September. While the parliamentarian troops were repulsed in some places, in others, the assaulting troops surged through the defences and Rupert, in the castle, found himself cut off from the rest of the city by Cromwell's cavalry. Offered generous terms, he quickly surrendered, leading his troops away to Oxford. Those who had previously persecuted Fiennes for giving up the city in 1643 realised that the same result had been forced on Rupert by similar circumstances. Charles, however, misled by Rupert's prediction of a siege lasting four months, was infuriated by his nephew's apparent treachery in surrendering Bristol with barely a fight and, suspecting him of plotting a coup, never really trusted him again and, issuing him with a passport, banished him to seek his 'subsistence, somewhere beyond seas'.

Gloucester in the Civil War

The city of Gloucester, with its bridge over the Severn, controlled the routes between the Southwest and the Midlands as well as that linking the king's capital of Oxford with royalist South Wales. Its situation in the middle of royalist territory rendered it both dangerous and vulnerable, particularly following the capture of Bristol to the south, and Cirencester to the south-east. Troops from Gloucester's Trained Bands were involved in the unsuccessful move on Worcester which ended at Powick Bridge in September 1642. By the spring of 1643, Welsh royalist forces from the Forest of Dean had advanced on Gloucester and 2,000 men camped at Highnam, just 2 miles (3 km) from the city. The parliamentarian General Waller and Colonel Massey, the governor of Gloucester, carried out a pincer movement and destroyed this force, killing 500, capturing 1,400, and demolishing Highnam House. However, with the fall of Bristol in July and the discouraging speed with which it had occurred, it became inevitable that the threat to Gloucester would increase. Massey therefore put every possible effort into improving the city's fortifications. The medieval walls remained around the eastern half of the city, 7 feet (2 metres) thick and around 10 feet (3 metres) high with an external ditch and fortified gates. These walls were strengthened internally with thick banks of earth to absorb shot and some of the gate-towers were packed with earth. In order to ensure that attacking troops might be kept as far away as possible, earthwork defences were added to the outside, based on the principles of continental fortification theory and practice but, of necessity in the time available, less robust than Massey would have liked. These additions consisted of a rampart and ditch fronting the city wall, with bastions added outside the major gates to provide flanking fire against any assault. The rampart was continued along the river frontage and also parallel to, and to the north of, the road from the West Gate to the bridge over the Severn. Dams were constructed to ensure that the waters of the Severn were exploited to the full in filling the moats but, unlike at Bristol there were few highpoints to afford dominance to the defenders. Whitefriars Barn on the eastern enceinte occupied the only commanding position which dominated that side of the city. It was strengthened by packing it with earth and armed with cannon to flank the walls to each side. When the defences of the south-eastern corner were threatened with being breached, a retrenchment of rampart and ditch would be dug diagonally between the East and South Gates. Unpopular though it was with the populace, some 241 houses in the suburbs outside the city walls were razed to improve fields of fire and to deny cover to attacking troops. This scorched earth approach was extended to the demolition of Llanthony Priory's tower, denying the enemy its use for observing the fall of shot and the selection of targets. During the course of the siege, further barricades were erected inside the East Gate and a sally-port was cut in its southern tower in order to disrupt the besiegers' mining operations. A small half-moon sconce was added to the new rampart near the castle on the quay, to counter royalist batteries across the river. The royalists' main bombardment, however, would take the East Gate as its primary focus. The bishop's palace of The Vineyard at Over had been refortified earlier in the year as an outwork to cover the northern approach to the city, especially prior to the fight at Highnam. It was captured, or rather occupied, by the royalists only after being abandoned by the defenders. The city's garrison consisted of around 1,400 soldiers and members of the Trained Bands, with some fifteen artillery pieces including at least one heavy gun. To provide the garrison with the necessary munitions, a gunpowder and saltpetre works was established in the Cathedral Precinct and Quay. Every ounce would be needed as the king had brought an army of over 10,000 to the fray with a large siege train with guns firing iron or stone cannonballs weighing up to 25 lb (11.3 kg) and mortars capable of dropping large explosive *granadoes* into the city.

After his success at Bristol, Charles I left a garrison behind and marched on Gloucester, arriving outside the walls on 10 August 1643 and demanding its immediate surrender. Over the next four weeks, the siege proceeded in an apparently uncoordinated fashion as the king had apparently been led to expect no more than token resistance and appeared unwilling, at that early stage in the war, to do too much damage to his loyal subjects and their property. Seeking to control the city's water supply, the king established his HQ at Matson House, built in 1575 and standing adjacent to the old reservoir on Robinswood Hill. Batteries were established against the South Gate on Gaudy Green, and at Llanthony Priory. This became a base for the royalists who used its buildings

Gloucester, Llanthony Priory, site of a royalist siege battery.

Gloucester, Matson House, headquarters of Charles I during the siege.

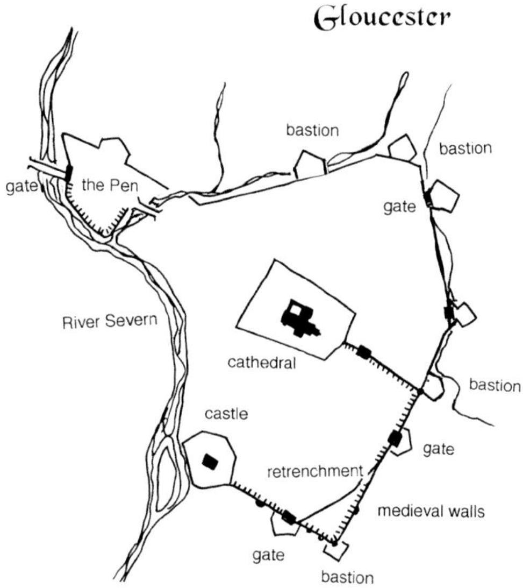

Sketch plan of the defences of Gloucester in the Civil War. The medieval defences of the city and the cathedral close were supplemented in 1643 by earthwork bastions linked by stretches of rampart and ditch. In the south-east corner a retrenchment was constructed behind the medieval wall. The Pen was defended and three of the gates on the eastern and southern walls were provided with outworks. After 1644, further bastions were added to the city's eastern defences.

as protection for a battery of heavy guns and was consequently damaged. Once his invitation to surrender had been rejected it would appear that the king was at a loss as to which of the available courses of action to follow: isolation and attrition leading to the city's starvation; a technical operation involving saps and approach works leading to a breach in the defences; or an immediate storm, utilising his overwhelming numbers of troops. Over the next month elements of all three alternatives would be seen. Dr Chillingworth reinvented ancient siege engines: four-storey wheeled towers whose cartwheels would lodge in the moat allowing drop-down drawbridges to reach onto the wall-head; mining operations, thwarted by underground springs, were carried out at the East Gate; batteries pounded the defences and fired red-hot shot to start fires; ditches were filled with faggots to facilitate an assault; and attempts were made to trick the defenders into issuing forth by pretending a relief force had arrived. The defenders were active in resisting the attackers by sallying out to spike the guns, to set fire to encampments and to throw grenades into the enemy siegeworks. On one occasion, no fewer than 400 men, a quarter of the active garrison, were committed to a sally at Kingsholm when they spiked one of the four cannon there. They ensured that the siege towers were unable to approach the ramparts by digging outer ditches which prevented the belfries' bridges from reaching the walls. The effects of the greatly feared mortars were particularly unpredictable. When *granadoes* exploded successfully, those on the receiving end, surrounded by the resultant carnage, were terrified and demoralised, but it was difficult to judge correctly the length of fuse. Too short a fuse and the *granado* exploded in flight, probably ineffectively; too long, and it might land without exploding, thus allowing defenders to render it harmless. One Gloucester resident was able to douse an unexploded 60-lb (27-kg) *granado* with a bucket of water as it landed at her feet. Unreliable artillery could be more of a danger to their operators than to their intended targets and one of the royalist mortars was recorded as exploding with its first discharge. For the duration of the siege, hovering in the minds of besiegers and besieged alike, was the awareness that a relief force might be on its way and, indeed, on 26 August, the earl of Essex left London with 20,000 men. After one last attempt to undermine the defences, whilst the defenders constructed more retrenchments, in the light of the imminent arrival of Essex, signalled by beacons, on the 5 September the king began the withdrawal of his army to Oxford via Stow-on-the-Wold.

The city had suffered a great deal of damage during the four-week siege but the clear-up was accompanied by efforts to improve the fortifications lest a further attempt be made by the royalists to complete the job. Over the next few years, bastions were added at more points on the circuit, particularly in the south-east and north-west sectors. Around Gloucester, outlying garrisons were established at Lypiatt Park, Eastington Manor, Prestbury Priory, Huntley, and Miserden Park. Lypiatt was taken in a surprise attack by Sir Jacob Astley in January 1645 but unwilling to remain in hostile territory, he withdrew, burning the house as he left. Westbury Court and its adjacent church with its detached tower were given a garrison by parliament in 1643 but the troops switched sides, requiring Colonel Massey to return the strongpoint to parliamentarian control in May 1644. Massey fought several campaigns not only across the Severn in the Forest but also in the areas around Tewkesbury and Cheltenham, while the royalists plundered Winchcombe early in 1645. He also entered into elaborate deceptions to entice the local royalists into an ambush within the walls of Gloucester but, despite efforts over several months, nothing came of them.

Gloucester's garrison remained loyal to parliament throughout the Second Civil War of 1648, but in 1650, in the Third Civil War, the city was required to spruce up its defences again in the face of an advancing royalist army whose vanguard, it was thought, was heading for the city. This force, surprisingly, was led by Massey who had thrown in his lot with Charles II. Massey had originally chosen to fight on the side of parliament because, as an experienced career soldier, he had foreseen more chances for promotion there. Although hailed as a hero for his defence of Gloucester and his subsequent successful campaigns, these opportunities for personal advancement failed to materialise and he became disillusioned enough to switch sides. Only after a decade of hair-raising adventures, did the Restoration finally allow him to fetch up back in Gloucester as its MP in 1660.

Cirencester in the Civil War

One of the techniques used in storming a town was to use fire-pikes to set fire to buildings on its periphery so that defenders were forced back by the smoke, thereby creating a gap in the defences. This tactic worked at Cirencester when it fell to the royalists on 2 February 1643, when Princes Rupert and Maurice had been quartered at Abbey House prior to sacking the town. Cirencester also suffered the terrifying experience of having bombs rained down on its buildings when a mortar was set up just 150 paces from the defences. After the surrender, further houses were deliberately set on fire, and it was said that the town suffered more casualties after its surrender to Rupert than during the actual fighting. Rupert then extracted £3,000 for the establishment of garrisons, plus a further £4,000 per month for their maintenance. The town was briefly retaken by Essex but soon relinquished and held for the king by Sir Jacob Astley. King Charles I stayed there for a night in 1644, but Fairfax and his colleagues received a warmer welcome when they passed through the town in 1646.

Tewkesbury in the Civil War

Colonel Massey from Gloucester captured Tewkesbury in April 1643 and the town gave Waller a place to fall back on after his defeat that month at Ripple, over the border in Worcestershire. After a further setback at Worcester, Waller once again withdrew to Tewkesbury but soon felt insecure there and abandoned it to the royalists, retreating to Gloucester, from whence he marched out taking a significant part of Massey's garrison with him. In the mid-summer of 1644, royalists from the garrison at Newent combined with Prince Rupert's force, threatening to cut parliamentarian troops off from their base in Gloucester, and setting up a new garrison in Tewkesbury. Massey took the unprecedented step of directly petitioning Parliament, as recorded in the State Papers, to have his troops returned as: 'it was with much difficulty we could hardly preserve our neighbours from ruin' (at the hands of Colonel Mynne in the Forest and Sir William Russell in Worcester, and additionally, against the King with his 4,000 foot, numerous cavalry, and twenty artillery pieces in the Tewkesbury area) as '... we have only two regiments of foot for keeping and maintaining Gloucester and Tewkesbury garrisons, besides four more small garrisons'.

However, despite his dwindling forces, the town was successfully stormed by Massey on 4 June. Later that month Waller gathered his forces together at Tewkesbury prior to his defeat at Cropredy Bridge near Banbury. By this time Tewkesbury had changed hands some eight times and King Charles later laid siege in the hope of securing a route into Wales but the royalists from Newent were defeated and the siege abandoned. Despite its strategic importance and the frequency of its changes of occupation, the town never had more than slight earthwork defences.

The Forest of Dean in the Civil War

Although owing no particular allegiance to either side, the inhabitants of the Forest of Dean found themselves at the centre of a number of skirmishes. In early 1643, Lord Herbert of Raglan Castle determined to counter the influence of the parliamentarian stronghold of Gloucester. Assembling a force of around 2,000 men, he marched into the Forest. Having scattered a newly raised parliamentarian regiment in Coleford, despite incurring significant casualties in the ensuing house-to-house fighting, Herbert marched on Gloucester. At Highnam, which he had fortified with earthworks, his force was routed and around 500 were killed attempting to flee the scene. Soon after the siege of Gloucester had been raised, skirmishing resulted in the destruction of the church at Taynton. It had been used to store powder and shot and was protected on three sides by defensive earthworks. For some reason, whichever side was forced to abandon it, set fire to it. In 1643, Westbury Court had been garrisoned by parliamentarian troops who defected to the royalist forces besieging Gloucester. In May 1644, the manor house and church at Westbury-upon-Severn, now held by royalist troops, was heavily damaged when it was stormed. Descriptions of the action relate how the parliamentarian assault troops placed ladders under the nave windows and threw *granadoes* inside, driving out the defenders. Another major skirmish took place in April 1643 at Littledean when Waller attempted to follow up his victory at Highnam by advancing into the Forest but, met by a formidable force of some 2,000 royalists under Prince Maurice, he was fortunate to extricate his troops. Maurice then set up his base at Dean Hall but, in May 1644, the royalist garrison there was taken by surprise by Colonel Massey's force. Around twenty were taken prisoner but two royalist officers were killed in the fighting. The next target for the clear-out of royalist garrisons was Newnham-on-Severn where defences had been thrown up to extend the castle earthworks in 1643. Again, the church became the centre of royalist defence and a contemporary, if partisan account, relates how a barrel of gunpowder was ignited inside the church in an attempt to discourage the attacking troops. This reckless act, apparently, was the cause of parliamentarian anger being vented on the remaining defenders, twenty of whom were killed and one hundred taken prisoner. In his absence from Lydney, 'Whitecross', the house of the Ironmaster, Sir John Winter, was successfully defended by his wife, but then destroyed by Sir John himself, to prevent it falling into parliamentarian hands.

The Beachley peninsula was regarded as the key to Wales and the royalists there were desperate to keep the passage across the Severn open in order to maintain contact with their forces in the south-western counties. In September 1644, Prince Rupert deployed a force of 500 cavalry to secure the passage, fortifying their positions and seeking to benefit from the guns of the royalist ships in the estuary. Colonel Massey attacked at low water, when those ships were out of range, and cleared the royalists out. However, they returned the next month under Sir John Winter, when Massey, once again, attacked, killing thirty men and taking 220 prisoners. Winter, who had made good his escape by scrambling down a cliff into a waiting boat, remained active in the area and left his base at Lydney in the spring of 1645 when he attempted to establish a river crossing at Lancaut, an attempt which would prove unsuccessful, as the parliamentarians had maintained a garrison at Tidenham throughout. Winter's legendary escapades were embellished over the centuries to give him hero status, but his contemporaries characterised him as a devious and self-seeking businessman and an unpopular and unsuccessful commander to boot.

Other Locations of Sieges, Battles, or Skirmishes during the Civil War

Lord Chandos having declared for the king, Sudeley Castle fell to Massey in January 1643 after a short siege but was quickly retaken by Rupert. At some time during the war, the Chandos household evacuated to Roel Farm at Hawling and the castle served as the royalist HQ in the north of the county and then underwent a long siege, only surrendering to Waller in June 1644. In 1646 orders were issued for the castle to be slighted, a process carried out in 1649 and it would be two centuries before reconstruction on any scale would take place. The ruined medieval Beverston Castle was garrisoned for the king in 1642, holding out against parliamentarian troops until May 1644, and then being taken only temporarily as two goes were needed to secure it. Berkeley Castle was initially occupied by parliamentarian troops in 1642, but Rupert's advance on Gloucester early in 1643 prompted them to withdraw and Rupert contributed 100 cavalry and dragoons to the garrison as it prepared to be besieged. These reinforcements secured the royalists use of the castle as a local HQ, and a prison for some of those captured at Bristol, until September 1645. The garrison of 500 men with eleven cannon under Sir Charles Lucas then spurned the opportunity to surrender the castle. The church had been integrated into the defences with its door being loopholed. After it had fallen to the attacking force with the loss of forty killed and ninety royalists taken prisoner, its roof provided the platform for cannon. The castle then suffered a three-day bombardment from close range, along with plunging fire from a mortar positioned on The Leys. Within three days, Lucas was granted terms and marched out with his men. The fall of Berkeley has been reported as a contributory factor to the king quitting Worcester and heading north, only to surrender to the Scots at Newark-upon-Trent. There were other royalist garrisons, especially during the siege of Gloucester, at Boddington, Brookthorpe, Hartpury Court, and Painswick's Court House. Wotton under Edge was the scene, in November 1643, of an attack by parliamentarian troops on a convoy of Irish troops travelling from their disembarkation at Bristol to join the king in Oxford. Heavily outnumbered, the attackers were beaten off. The fortified

Berkeley Church, whose detached tower was sited at a distance from the castle to prevent its use by an attacker, nevertheless sustained damage during the siege of 1645.

house of Yate Court was razed in 1644 after occupation by parliamentarian forces.

The Battle of Stow-on-the-Wold, one of the last major actions of the First Civil War, took place on 21 March 1646 after Sir Jacob Astley had gathered the nucleus of a royalist field army together from the remaining garrisons in the West Midlands. Although Chester fell before he was able to prevent it, he was hoping to reach the beleaguered king, so he headed toward Oxford with his force of about 3,000 men hoping to raise the siege there. Marching his army through the Cotswolds he was intercepted by a strong parliamentarian force under Sir Thomas Morgan, governor of Gloucester, with another large force under John Birch, governor of Hereford, on his flank. Coming up behind the royalist army was Brereton's parliamentarian force, now liberated from Chester. Moreover, between Oxford and Stow, a fourth parliamentarian force was stationed to prevent any relief reaching the king. Despite making attempts to outmanoeuvre his opponents, Astley became boxed in and many of his men broke and fled. The royalist army lost 1,600 prisoners and ceased to exist as a force. After the battle, many of those royalist prisoners were housed in St Edward's church in Stow.

Munitions

William Winter was vice-admiral of England and had been granted the manor of Lydney and his son Edward had developed the charcoal and iron industry into a lucrative family business, while cutting across the Forest-dwellers' commoners' rights. Edward's royalist son, Sir John, founded Gunns Mills at Flaxley (SO675159) as a blast furnace and cannon foundry in 1625. In 1629, Charles I's master of the ordnance placed an order for 610 guns to be made and sent to Holland, some of them being used by each side in the Civil War. As a direct result of his military activities, Sir John was committed to the Tower of London in 1644 and the site was ordered destroyed by parliament after the war in 1650 but rebuilt in 1682. The scant remains of this and later structures subsequently used as paper mills were surveyed by Cotswold Archaeology in 2016. Soudley Furnace supplied 500 cannonballs for the siege of Gloucester, and Thornbury was the location of a gunpowder works during the seventeenth century. Stroud mills, Ham Mill in particular, produced red cloth, operating under letters of protection from Princes Maurice and Rupert, while Charles I also wrote to his nephews authorising them to commandeer all the locally produced cloth for use by the royalist armies. Uley Blue and Stroudwater Scarlet were famous throughout Europe.

Restoration, Recovery, and Revolution

The war may have apparently come to an end, but military campaigning continued. A regiment had been raised for Cromwell's Irish Campaign in 1646 employing 180 men impressed from Gloucestershire. Colonel Morgan's regiment remained in Gloucester while Colonel Blount's would serve in Ireland and Massey was named as lieutenant-general of horse for the Irish adventure. This force was delayed and would eventually leave in 1649 under the command of Cromwell and would contain 400 impressed Gloucestershire men. In 1648, the cities of Bristol and Gloucester and the county of Gloucestershire were included in a list of seventy towns, cities, and shires with existing militia forces from whom 10,000 men were impressed for service in Ireland in 1651. The Restoration in 1660 then brought all the militia directly under royal control alongside a new standing army under Monck, now duke of Albemarle. In 1677, the Common Council of the City of Bristol secured permission to re-establish an Association of the Artillery Yard which had lapsed after the city's royalist occupation in 1642. This elite corps would practise on St Philip's Marsh and would demand more of a commitment from its members than that expected under the militia regulations.

Many places which had suffered damage during the Civil Wars took many years to recover from the effects on their social, economic, and structural fabric. After the final royalist defeat at the Battle of Worcester in 1651, the mayor of Gloucester put in a claim for compensation which was partially met by grants of confiscated Irish lands, but these had all been sold by 1679 accruing little profit. Much of the city's defences had been demolished during the time of the Commonwealth but the final order to raze the city walls was made by Charles II in 1662. Many of those who had been community leaders during the wars were dismissed or side-lined while royalist sympathisers were appointed to oversee the city's humbling.

When the duke of Monmouth, illegitimate son of Charles I, landed at Lyme Regis in June 1685 to depose his uncle, James II, and claim what he saw as his rightful inheritance as a Protestant prince, he had intended to march on Bristol and Gloucester; link up with supporters coming down from Cheshire; and then advance on London at the head of a massive army of followers. Ignoring advice, he rejected an advance on vulnerable Exeter and headed directly north for Bristol. Here he was faced by the Gloucestershire Militia, four regiments of foot and one of horse, under the duke of Beaufort defending the city, and a force of 2,000 regular troops at Bath. Although he could probably have taken Bristol which had quickly been secured for the king by the City of Bristol Militia, he passed up the opportunity and chose to focus on Bath. Despite having greater numbers and winning an initial engagement at Norton St Philip, Monmouth's scantly armed and untrained peasantry would be no match for the professionals of James's regular army who would decisively defeat him at Sedgemoor on 5 July 1685.

The Development of Gloucestershire's Military Institutions 1660–1815

The Gloucestershire Regiment

The earliest regiments regarded as belonging to the British Army were associated with the people who raised them rather than the locations from which they had been recruited. The 9th Foot was raised in Gloucester in 1685 as Henry Cornewall's Regiment of Foot, and in the next century would often be based in the county and would recruit nearly 200 men from the North Gloucestershire Militia. However, in 1782, when George III ordered regiments to be linked to specified counties for their depots and recruitment, the Ninth became the Norfolk Regiment. The origins of the Glosters lay with the 28th Foot and the 61st Foot, raised, respectively in 1694 in Portsmouth, and in 1756 from a battalion detached from the 3rd Foot during the Seven Years' War. In the 1782 reallocation, they became the 28th (North Gloucestershire) and the 61st (South Gloucestershire) Regiments of Foot. In the early years of the French Wars, the 61st were in the West Indies and then joined the 28th in Egypt. Both regiments then fought in the Peninsula, while the 28th went on to fight at Waterloo.

The Gloucestershire Militia

By 1697, the Gloucestershire Militia consisted of the White, Green, and Blue Foot Regiments based on Gloucester with a total of twenty-nine companies containing 1,668 men; the Red Regiment at St Briavels with a further 531 men in nine companies; and six troops of cavalry, with a strength of 243 troopers. Ten companies with a total of 727 men formed the City of Bristol Militia. Following legislation (George II, Act 30, 1757), which revitalised what had become a moribund force, a combined County Militia regiment was formed in 1759 of two battalions, North and South, calling for men between the ages of eighteen and fifty, chosen by ballot, to serve for three years. The regiment's establishment was set at 960 drawn from Gloucester and Bristol, the cities and shires. The militia was intended to provide internal security and support for the Regular Army in wartime. At the end of the Seven Years' War in 1763, the regiment separated into two: South Gloucestershire at Gloucester and North Gloucestershire at Cirencester. Both regiments were called out in 1793 for the French wars and given their 'Royal' prefix two years later. Later legislation (George III, Act 42, 1802) called for Gloucester and Bristol to raise 1,163 men. Following the swings in the fortunes of the wars both regiments saw intermittent service. The Royal North Gloucestershire Militia was serving in Ireland in 1798 and then as garrison troops in Dover Castle, and back to Ireland in 1813–14. The Royal South Gloucestershire Militia carried out anti-invasion duties in barracks at Horsham (Sussex) in 1798 and served in Ireland, 1815–16.

The Gloucestershire Yeomanry

One reaction to fears raised in England over the French Revolution was the appearance of troops of yeomanry. These were often raised by members of the aristocracy and other landowners, drawing on their tenant farmers and neighbours. Once it became apparent that war with France was inevitable, under the 1794 Volunteer Act, lords lieutenant were charged with raising troops of volunteer cavalry for the nation's defence against invasion. The county's first such troop was raised in Cheltenham in 1795, soon to be followed by troops in Wotton-under-Edge and Minchinhampton and, within three years, there were eight troops in existence. That at Gloucester met in the Bell Inn and further troops were raised in Stow-on-the-Wold, Bristol, and Henbury. Yeomen supplied their own horses; uniforms were provided by their generally wealthy leaders; and weapons came from the government. After a short hiatus caused by the Peace of Amiens in 1802, a renewal of the invasion threat saw the county's yeomanry troops increase to thirteen, including the Bristol Volunteer Cavalry; the duke of Beaufort's Badminton Troop; and the Stow Volunteer Cavalry. The Yeomanry also served a function in internal security, quelling riots and brawls, and performing ceremonial duties, escorting the future George IV in Gloucester in 1807. After Waterloo in 1815, all but the Gloucester troop, retained until 1827 for policing duties, were disbanded.

The Volunteers

Since many citizens who were unable to join the Yeomanry wanted to do their bit or, as the cynical would have it, wanted to avoid the Militia Ballot, corps of volunteer infantry such as the Loyal Westbury Volunteer Corps and the Severn Rifle Corps, as well as Artillery were also raised. In 1797, the Bristol Regiment of Volunteer Infantry with an establishment of an adjutant, twenty-two NCOs, and 1,000 men, joined the Bristol Garrison along with three regiments of militia, one of them from Buckinghamshire, and some volunteer cavalry. Not before time as it transpired that the French force which landed at Fishguard had been heading for Bristol with the intention of destroying shipping in the port. The Suffolk Fencible Cavalry were despatched to the coast to repel any attempted invasion. The Loyal Pill and Port of Bristol Volunteer Association had been formed to man guns both on land and afloat but accomplished little before becoming the Pill Sea Fencibles. Another local volunteer artillery unit was raised in Berkeley. In all, some three dozen volunteer infantry units were raised from 1803 onwards.

Preparations for War: Logistics, Recruitment, and Deployment

Continuing the counties' traditions of munitions production, in 1749 there were gunpowder warehouses at Westbury-on-Trym and in 1775, William Watts, a plumber, pioneered a new way of making lead shot. He converted his house on the corner of Redcliffe Hill and Redcliffe Parade in Bristol into a manufactory by adding a three-storey tower on top of his three-storey house and excavating a shaft underneath the house, giving a total drop of 90 feet (27.5 metres). At the top of the tower, molten lead was poured through a zinc colander to fall into cold water at the bottom of the shaft. Unknown to Watts, it was the arsenic in his lead supply which enabled the process to work. The Shot Tower was operated until 1968 when it was declared unsafe and demolished, to be replaced by the Cheese Lane shot tower on the north bank of the floating harbour, now a listed building. Gloucestershire mills had continued production of blue and the red cloth for uniforms throughout the Napoleonic Wars, and in Stroud, Lodgemore Mill produced the distinctive scarlet cloth used for the uniforms of the Foot Guards. Owing to the country's generally poor roads, troop movements were slow and uncomfortable and with an expected French invasion in 1803–04, it was imperative that measures to respond to an enemy landing were urgently put in place. One solution was to make more use of the new and expanding canal network and once the important connections between London and Liverpool were established, the War Office looked to establish further routes. Sir Brook Watson, the commissary general, therefore identified canal routes linking the Thames at Lechlade to the Severn at Gloucester via Cirencester and to Framilode via Stroud, recording both the canal transport available and the supply of horses and their changeover points.

Gloucester's gates were demolished in the 1780s, and in 1791, a prison was built on the site of the castle. A customs post was set up on Beachley Point to inspect ships sailing up-river to Bristol and the Revenue Men were billeted in a cottage in the village. In 1793, a dispute over the continued collection of tolls on Bristol Bridge saw riots which resulted in troops of the Herefordshire Militia being called out to support the magistrates. A series of volleys and a bayonet charge caused eleven deaths and forty-five serious injuries. Militia units were generally employed outside their own areas to avoid them being subjected to local pressures but a suggestion that the Herefordshire Militia might return to the city in 1797 was strongly resisted by the mayor, with food riots in 1803 being suppressed by the 1,000-strong Bristol Volunteer infantry. The coming of war with France saw the establishment of a naval Impress Service office on Bristol's College Green, but recruitment was difficult, and parliament was forced to pass a law requiring merchant shipping to provide seamen for the navy prior to being permitted to sail, eventually offering generous bounty payments to soften the blow. The Impress Service in Bristol was allocated three tenders for boarding ships entering and leaving the port. While recruitment into the new volunteer corps was slow in Bristol, a 700-strong regular unit, the Loyal Bristol Regiment or 103rd Foot was recruited and deployed to Dublin, ultimately to become the Royal Dublin Fusiliers. Another local volunteer infantry unit was raised in the middle-class areas of Clifton and Westbury-on-Trym.

Measures taken to guard against invasion included the stationing of two twelve-gun, 150-ton gunboats in the Severn Estuary and a chain of signal stations straddling the mouth of the Bristol Channel. Local shipowners were enjoined to arm any vessel capable of carrying a small cannon to provide a force of gunboats capable of repulsing an attack up the Severn. Three 24-pounder cannon were to be mounted in a permanent battery on Portishead Point but in the interim, the mayor of Bristol organised temporary batteries, each equipped with four captured French 36-pounders and manned by the Invalid Artillery or Severn District Sea Fencibles at Portishead Point and King Road Farm, Avonmouth. Gunners manning the latter battery would later be accommodated in Barracks Cottages, marked on the 1st Edition of the OS map, and the Royal Bristol Artillery Volunteers may have helped man these guns. The official Ordnance Board records for this period list the four 36-pounders at Portishead Point and those four more at Avonmouth. The barrack block at Portishead could accommodate an officer and forty-five men. Stapleton Prison had been built in 1779 by the Admiralty to house American naval prisoners landed at Bristol during the American War of Independence, most of whom had been released by 1783. It then housed French officer POWs from 1793 to 1813, being enlarged in 1804. After 1815, the buildings were used as an arsenal and a nautical school for wayward boys and, after 1834 as a workhouse. The Militia Garrison Guard House and mess survived in the former Workhouse, as the Blackberry Hill Hospital, now a housing development. As part of a national barrack building programme, barracks accommodating three troops of cavalry, 158 troopers and their officers, were proposed for both Bristol and Gloucester but never built. However, in response to Bristol's innkeepers' pleas for relief from billeting troops, the War Office leased the Royal York Terrace in Clifton for use as a barracks, but this development coincided with the Peace of Amiens in 1803 and was not advanced, the lease being sold on, at a profit, in 1809. After the wars, Bristol's garrison continued under the previous unsatisfactory billeting arrangements. One unfortunate consequence of wars is the proliferation of weapons and their indiscriminate use. In Gloucester in 1802, a fifteen-year-old discharged soldier used his musket to threaten passers-by on Bristol Bridge and finally shot and killed a man.

Civil Disorder

The period leading up to the Reform Act of 1832 saw social upheaval in the counties. As the rise of industry unbalanced the social mix and saw the dramatic growth of the towns, so opposition to the introduction of machinery and the disruption to lives rooted on the land and in cottage industries caused violent scenes. The Stroudwater Riots of 1825 followed the introduction of the Flying Shuttle into the mills, halving the number of weavers' jobs. The army was called in, soldiers were attacked, and looms were burned. The Forest of Dean also saw riots in 1831, led by Warren James, an impoverished miner, voicing the discontent of miners and metalworkers who had been deprived of their customary rights and denied access to large parts of the Forest, in the process of being fenced for the protection of timber being grown for the Royal Naval dockyards. When the crown repudiated the ancient forest laws, a process started in the 1770s, many were forced into either becoming the waged employees of outside industrialists, or retaining independence but forced into competing under impossible conditions. A petition to parliament failed, resulting in militant action. Up to 3,000 local men were involved in actions which included attacks on St Briavels Castle, still in use as a jail, and the military were brought in to quell the disturbances. James, along with others, was charged with felony and sentenced to death but this was commuted to transportation to Tasmania where he died ten years later.

Alongside industrial and social unrest came political turmoil. Bristol was one of the cities where violent riots followed a second failure of the Reform Bill in 1831. The arrival of the Recorder for the 1831 Assizes was marked by excessive security. Two troops of light dragoons, the 14th Hussars, were quartered in Clifton with another at Keynsham, and 300 constables were mobilised. A large crowd had gathered to display their dissatisfaction with the Establishment and the Riot Act was read. Very soon, fighting erupted with the constables and the troops being called in to disperse the rioters who had stormed the Mansion House and the Council House. It seems likely that the civil and military authorities, the leaders of the Bristol Political Union (BPU) and the thugs sworn in as special constables all contributed to the mismanagement of an already volatile situation. Once rioting was under way, it was dealt with in a heavy-handed way by the specials and by some of the military. The jails were emptied, the civic buildings and the bishop of Bristol's palace were torched, the bishop seeking refuge in Knole Park, and damage, assessed at £300,000 (£30 million in today's values) was caused. At least twelve citizens were shot by the soldiers and many more bodies were found in the burnt-out buildings. A recent historian has set the death-toll at between 120 and 250. Thirty-one of the 102 rioters who were tried were sentenced to death with four of those sentences being carried out along with twenty-two who were transported. Things could have been handled much better, and some of those held responsible were brought to account. The military commander, Colonel Brereton, was court-martialled and committed suicide before a verdict could be reached, and the mayor was arraigned for negligence but acquitted. The BPU had offered to intervene but had been rebuffed; the Tetbury Troop of yeomanry cavalry spent a week in the city but a forty-strong Dodington Troop, which had arrived to help, was sent home as there was no coordination of effort by army officers unable to cope with the chaotic situation. A belated appearance by the Bedminster Troop did little to ease tensions and only the appearance of the regular cavalry from Keynsham and Gloucester finally restored the peace. One of the problems causing the escalation of violence and destruction had been the scarcity of either *bona fide* constables, acceptable to the community, or volunteers, but it would be a while before steps were taken to address this deficiency. While Cheltenham had set up a police force in 1831, it was only joined by Gloucester in 1836, and by Cirencester in 1839 when a county constabulary was formed. Bristol's force was inaugurated in 1836 to the requirements of the Municipal Corporations Act, with 232 officers based in four divisional police stations.

4

Victorian and Edwardian Gloucestershire 1840–1914

Alongside all this turmoil, infrastructure was developing apace: the development of Bristol's Floating Harbour had commenced in 1803 with the 'New Cut'; the Gloucester-Sharpness Canal had been dug by 1827; dockside warehouses and a customs house were built in Gloucester in 1845; new docks would open at Avonmouth in 1877 and at Portishead two years later; between 1840 and 1878 the Great Western Railway built tunnels, bridges and railway stations, many to the designs of Brunel; the Severn railway bridge dates from 1879 and the Tunnel from 1886; and the line from Birmingham to Newport and South Wales passed through Cheltenham and Gloucester.

The Regular Army, the Militia, and the Volunteers in the Nineteenth Century

The Regular Army

The 28th Foot, or North Gloucestershire Regiment, spent the decades after Waterloo in the Mediterranean, Ireland, and Australia. It served in the Crimean War, taking part in the battles of Alma and Inkerman and the siege of Sevastopol. The 61st Foot, or South Gloucestershire Regiment, served in India during the Indian Mutiny or First Indian War of Independence. Horfield Barracks, Bristol, was built as a response to the prospect of Chartist unrest with the memory of the riots of 1831 in mind. It was reputedly opened by the duke of Wellington in 1847, the date over the former entrance archway, but the local press has asserted that there survives no record of such an event when the duke, though still titular C-in-C of the British Army, would have been nearing his eightieth year. Two other barracks built at this time in the region for similar reasons and in a similar style, were Newport, scene of the Rebecca riots, and Brecon, a focus for anti-Turnpike action.

The Cardwell and Childers reforms of the last decades of the nineteenth century saw the establishment, in 1872, of locally based regimental districts. The 28th and the 61st Foot were brought together as a two-battalion formation in the 37th Regimental District. The intention was that this unit would provide one battalion for home and another for foreign service in rotation. In 1881, further reforms combined them as the Gloucestershire Regiment in the 28th District. Unusually, rather than occupying a new build, the 37th (and later, the 28th) District was based on the already existing Horfield Barracks. The barracks was laid out as a quadrangle around a barrack square, all surrounded by defensible, stone, loopholed perimeter walls with projecting bastions at the four corners and entered through a gateway flanked by sentry posts. Along the north side stood three long terraces of two-storey barrack rooms. The west range, facing onto the main road, consisted of an impressive block which, on the evidence of other contemporary barracks, contained the guardhouse, armoury, sergeants' quarters and orderly room. The east range was a long, two-storey block of officers' quarters and mess, with the CO's house standing in front of it. Ranged around the perimeter walls were stables, stores, workshops, hospital, magazines, washhouses, cookhouse, forage barn and privies. Outside the barrack perimeter stands the Garrison Chapel, designed in 1857 by Butterfield, now deconsecrated and converted into offices.

Bristol, the main gate of Horfield Barracks with its defensible perimeter wall.

The Militia

Following the end of the French wars, the militia fell into abeyance, but reinvigorating legislation was passed in 1852 and, a year later, further legislation required the construction of militia armouries. Across the country, many of these were built as toy forts and the pair built in Gloucestershire were amongst the finest examples of this genre. The near identical armouries of the 69th Royal North Gloucestershire Regiment of Militia in Cirencester and of the 23rd Royal South Gloucester Light Infantry Regiment of Militia in Gloucester, were both designed by local architects Fulljames & Waller in 1857 and were constructed in local stone around a barrack square with an imposing front range with turrets, battlements, and arrow-slits, containing the armoury, mess, and orderly room. Ranged around the square were stores, workshops, and accommodation for the small permanent staff. The Gloucester armoury stood to the north of the prison and was largely demolished in the 1960s. The Cirencester armoury on Cecily Hill still stands. The militia had traditionally recruited by ballot but now attempted to attract volunteers who might be more amenable. Militiamen trained for six months and were then liable to be called out to serve in national emergencies or could volunteer, as many did, for the regulars.

In order to free up regulars for service abroad, the two militia regiments both served during the Crimean War when over 600 men of the Royal North Gloucestershire Militia volunteered to serve in Ireland. In 1881, as a corollary to the formation of the Gloucestershire Regiment, the South and North regiments

Cirencester, the Cecily Hill Militia Barracks of 1857.

became, respectively, the 3rd and 4th (Militia) Battalions of the Gloucestershire Regiment, based at Horfield Barracks training recruits for the two regular battalions and maintaining the Reserves of time-expired men. The South African War saw both battalions providing volunteers including a combined force which went to St Helena in the south Atlantic as guards for the Boer POWs.

The Rifle Volunteer Corps

As a result of the Regular Army's performance in the two major campaigns of the 1850s in India and in the Crimea, concern was widely expressed over the leadership qualities of the officer class; the competence of the Commissariat and medical services; and the martial skills of the troops themselves. This disquiet was exacerbated by disquiet over French ambitions under the latest Napoleon, Bonarparte's nephew, Britain's ally in the Crimea, but now seen to be pursuing expansionist policies in Europe including an overt challenge to Britain's dominance in the Channel. We have seen how the French Revolution and the wars with Napoleon I, spawned a volunteer movement from which only some of the county yeomanry regiments survived. Attempts to raise new volunteer corps, in the guise of drill or rifle clubs, had been made earlier, in response to Napoleon III's seizure of power. In 1852, Sir George Grey, the home secretary, while resisting wholesale requests to raise volunteer units, sanctioned just two, including one, based on a rifle club in Cheltenham, which appears not to have materialised in the end. Thus, it was not until 1859 that the National Rifle Association, a national, civilian initiative to promote shooting, followed closely by the formation of the military Rifle Volunteer Corps (RVC), achieved through parallel local efforts, brought all those efforts to fruition. While the Yeomanry tended to be a rural movement, taking its lead from the aristocratic landowners who enrolled their tenantry, rural squirearchy, and yeoman farmers, the RVC was an urban movement, promoted largely by the professional, merchant, and artisan classes. Thornbury's detachment would be led by the local solicitor. Public meetings were held, one of the earliest being called in Bristol in February 1859 to explore the possibilities of a local corps. This meeting was clearly successful as the City of Bristol RVC appointed its first officer that September and, by the following June comprised the basis of a full battalion of ten companies which in 1880 would become the 1st Gloucestershire RVC. The county's 2nd Corps formed in Gloucester Docks in October 1859, quickly followed by fourteen further corps across the county, not all of which, being subject to disbandment or amalgamation, survived. Enough, however, stayed the course and, in 1880, formed the ten companies, lettered A–K, of the 2nd Gloucestershire RVC.

Artillery and Engineer Volunteers

Alongside rifle corps, there were specialist artillery and engineer corps, reflecting the needs of a modernising army. The Gloucestershire Artillery Volunteers formed at the Whiteladies artillery ground in Bristol, receiving four 18-pounder smoothbore guns from Woolwich in April 1860. Further batteries were quickly formed in Gloucester, Newnham-upon-Severn, in the Forest of Dean, and in Somerset. After 1880, the corps had expanded to thirteen batteries spread across Gloucestershire and Somerset with a cadet company in Bristol. In 1893, the Artillery Volunteer Corps was assigned as part of any wartime Plymouth garrison and in 1902 was absorbed into the Royal Garrison Artillery (RGA). These local RGA units, along with the 400-strong Royal Naval Artillery Volunteers who operated in Bristol from 1874 to 1892, shared responsibilities for manning the Severn forts of Lavernock Point, Flat Holm, Steep Holm, and Brean Down with a total of thirty 7-inch RML guns, to block the Bristol Channel against sea-borne attacks from the west. The naval volunteers were based aboard HMS *Daedalus*, a dismasted frigate of 1826, berthed below Clifton.

The 1st Administrative Battalion of the Gloucestershire Engineer Volunteer Corps (EVC), seventh in the national order of precedence of such units, emerged in 1867 in Bristol as a combination of the 1st and 2nd Corps formed in 1861 in, respectively, Gloucester and Bristol. The 1880 reorganisation saw a new 1st Corps, with its HQ and four companies at Gloucester and outliers in Devon and Somerset, but the Bristol EVC with HQ at Clifton College had declined to join this new unit. In 1886, two new companies of Submarine Miners, responsible for defensive minefields protecting access to ports, were formed to operate in the Severn Estuary and in July 1889 the HQ of the 1st Gloucestershire EVC moved from Gloucester to Cheltenham, with an affiliated cadet corps at Cheltenham College. The cadet corps at Clifton College, meanwhile, elected to remain with the Bristol EVC.

The Yeomanry and the Light Horse Volunteers

The Gloucestershire Hussars (RGH) Yeomanry (Royal from 1841) was formed in 1834 from those residual Troops which had survived from the French Wars and taken part in actions in support of the Civil Power. Its CO was the 7th duke of Beaufort who had experience as an officer in the 10th Hussars. Membership consisted predominantly of local landed families, their tenants and farming neighbours. Uniform was the exotic hussar dress of Hungarian origin and obligatory moustachios. Yeomen drilled in small groups in rural settings; came together as Squadrons for ceremonial occasions and field days and might only parade as an entire regiment for annual camps at the Beauforts' Badminton Park, Cirencester Park, Sudeley Castle, Piercefield Park, Chepstow, Cheddar, and other similar venues. In 1835, seven troops were listed as dispersed across the counties, meeting for drills at venues such as Berkeley Castle, Oldown House, Tockington, and Horfield Barracks. Over six decades, troops were reorganised, amalgamated, and reformed.

The Yeomanry was widely regarded as being expensive and socially exclusive, so a unit of Light Horse Volunteers was formed in Stroud in 1864 as an alternative. Although some of the perceived disadvantages of belonging to the Yeomanry were overcome, membership was still expensive and sustaining a horsed unit was extremely costly for both individual members and their patrons. The Gloucestershire Light Horse lasted barely three years, disbanding in 1867 and selling their elaborate helmets to the 2nd Devonshire Light Horse who, themselves, got only ten years' wear out of them. By 1880, all but a handful of the forty-odd light horse and mounted rifle units across Britain had succumbed to these universal problems, ironically to be later reinvented as the Imperial Yeomanry of the Second Boer War.

By 1893, a new officially imposed Yeomanry structure had stabilised with four squadrons based in Gloucester, Badminton, Newport (Monmouthshire), and Bristol in 1899. Although drills continued to be designed to develop horsemanship through activities such as tent-pegging, swordsmanship and firing carbines from the saddle, this skillset was rapidly becoming redundant in modern warfare. Even reconnaissance, the last bastion of the cavalry's *raison d'être*, would be handed over to the balloons of the Royal Engineers and the string-bags of the Royal Flying Corps before too long. However, the Boer commandos demonstrated the value of combining the mobility of cavalry with the sharp-shooting skills of trained infantry, and Lord Roberts, commanding in South Africa in 1900, requested the formation of large numbers of mounted infantry to counter them. Although the Yeomanry regiments could not legally be deployed abroad *en bloc*, large numbers of their members volunteered for the Imperial Yeomanry, specifically formed for service in South Africa. In 1900, 125 yeomen of the RGH were trained as mounted infantry at Horfield Barracks, their new role being underlined by their designation, not as a squadron of cavalry, but as No. 3 Company, 3rd Battalion, Imperial Yeomanry.

Life in the Volunteer Force

From 1859, Volunteers were required to drill regularly, to achieve and maintain proficiency in shooting, and to sustain a soldierly turnout in the smart uniforms (green tunics in Bristol and grey in Gloucester) paid for by public subscription; a wealthy CO or public-spirited patron; or by a proud municipality. Early in their history, the War Office had decided to exercise control over what they had hitherto regarded as private armies by offering financial incentives. Corps would be assessed on the basis of the number of 'effectives' they were able to muster, that is men who met the criteria for turn-out, attendance at training, drills, and shooting proficiency. Following official inspections, grants for uniforms, equipment, and the costs of full-time instructors would be paid to successful corps out of public funds. One stipulation was that corps should provide a secure armoury for the storage of weapons and ammunition; an orderly room or office for the corps' administrative business; and an indoor space for drilling throughout the seasons. Premises, often combining those three functions, became drill halls. These could be purpose-built as at Queens Road and Old Market Street, Bristol; Stroud's Bowling Green; Cheltenham's North Street and the REs' hall on Swindon Road; or the Tewkesbury drill hall built by public subscription. In Tewkesbury, at the time of the Boer War, a public subscription raised £80 to purchase an old malt house in Red Lane and spend £520 converting it into a drill hall for 'D' Company of the Glosters' 2nd Volunteer Battalion. The hall measured 72 by 33 feet (22 by 10 metres) plus an armoury, clubroom, offices, and stores. There were other conversions of existing buildings: chapels at St George, Bristol, Newnham-upon-Severn, and at Dursley; almshouses at Winchcombe; skating rinks at St Michael's Square, Gloucester, and Park Row, Bristol; a large house and garden at St Michael's Hill; and the terraced 'Glenavon' in Daisy Road, both in Bristol. As the militia was taken into the mainstream Regular Army, their armouries at Gloucester and Cirencester became available for the Volunteers. Other venues might be shared; Coleford's town hall; masonic halls in Dursley and Stow-on-the-Wold; public houses with skittle alleys at Minchinhampton and Tetbury; and the public Bingham Hall in Cirencester with its rifle range and armoury.

Volunteers committed to regular drills, field days when isolated rural units might have the opportunity to combine in larger formations, and summer camps. Shooting would take place in indoor 25-yard ranges using either small-bore rifles or Morris Tubes which would adapt full-bore rifles for indoor use. Outdoors, shooting took place on 600-yard ranges such as existed at many locations including Drakestone Point, Stinchcombe, from *c.* 1880, and at Maugursbury.

Bristol, St George, the drill hall in use in 1910.

Tewkesbury, the remodelled drill hall on Back Lane.

Cirencester, Bingham Hall in King Street opened in 1908 as a public hall and rifle range; served as a drill hall for the Boys' Brigade and Church Lads' Brigade; was a Red Cross Auxiliary Hospital in the First World War; a recruiting centre in 1939; and was requisitioned by the War Office as accommodation in 1940.

Newnham-upon-Severn, the Dean Road drill hall occupied a chapel of 1826 from 1860 but had ceased operation by 1914.

Newnham-upon-Severn, the Severnside drill hall, marked on a map of 1903, had been a warehouse of the 1760s on the wharf in Back Lane but was no longer in use by 1914.

Cheltenham, the Swindon Road drill hall of the Volunteer Royal Engineers.

At Coney Hill Coberley, the Seven Springs rifle range west of Chatcombe Wood was in use by 1897, and a range at Sneedham's Green, Upton St Leonard's was up and running in 1905. Manoeuvres and camps were held across the county at such sites as Littledean Camp or Stinchcombe Hill.

The 1881 reforms which saw the creation of the Gloucestershire Regiment with its two regular and two militia battalions also brought the RVC into the regimental family. The Bristol-based corps became the 1st Volunteer Battalion (VB) and the Gloucester-based corps, the 2nd VB. By 1888 the 1st VB Gloucestershire Regiment was part of the Severn Brigade charged with the defence of the Severn ports, but a further reorganisation in 1906 assigned both of the Glosters' VBs to the Portland Brigade, defending the naval base of Weymouth and Portland Harbour. In 1900, a 3rd VB was formed in Bristol. Towards the end of the century, cadet units operated at Cirencester School, Bristol GS, and the Gloucester County School at Hempstead.

The Territorial Force (TF)

In 1908, the government followed the advice of Viscount Haldane and established a new volunteer force for home defence. This would be called the Territorial Force (TF) and would absorb all those volunteer units which opted to join. It would be centrally funded entirely by the government and administered locally by County Territorial Associations (CTA). In effect it would become a parallel army modelled on the composition of the regulars. Its fourteen regional divisions would each consist of four infantry brigades with each of these brigades containing four battalions, previously designated Volunteer Battalions but now numbered into the parent regiment. The Glosters' two militia battalions having combined as the 3rd Battalion, responsible for running the regimental depot and supplying drafts to the two regular battalions, the three Gloucestershire VBs became the 4th, 5th, and 6th Battalions. The 4th and 6th Battalions of the Glosters and the 7th and 8th Battalions of the Worcestershire Regiment formed one of the South Midland Division's four infantry brigades with its HQ at Charlecote, Battledown, Cheltenham. The Glosters' 5th Battalion joined with battalions from Oxfordshire, Buckinghamshire and Berkshire as the South Midland Infantry Brigade in the same division. Each TF division had an attached mounted brigade of three Yeomanry regiments. By now, the four squadrons of the RGH were settled in Gloucester, Stroud, Newport, and Bristol. The mounted brigade belonging to the South Midland Division included the RGH, the Warwickshire and Worcestershire Yeomanries, and the Warwickshire RHA. The 1st South Midland Brigade, RFA, consisted of three Gloucestershire batteries and their associated ammunition column based in Gloucester and Clifton. The Severn Division of the Submarine Miners had been converted to RE Fortress Engineers based in Cardiff in 1905, and only a single company of the Gloucestershire volunteer engineers joined the TF and then as part of the ASC. The HQ and three companies of the newly formed South Midland divisional REs were based in Park Row, Bristol, along with one section of the South Midland Divisional Signals Company, RE (TF). The ASC Transport and Supply Company of the Gloucestershire and Worcestershire Infantry Brigade was based at Wallbridge, Stroud. The HQ and the three sections of the 3rd South Midland Divisional Field Ambulance occupied Colston Fort in Kingsdown, close to the 2nd Southern General Hospital, RAMC (TF), then as now, the Bristol Royal Infirmary. Elements of the various volunteer engineer units transferred to the TF as Divisional REs. Clifton College's four engineer companies joined the Junior OTC, as did Cheltenham College's four infantry companies and the single infantry company at Bristol GS. Bristol University, chartered in 1909, formed a single volunteer infantry company in 1910 which joined the Senior OTC.

The CTA, lodged at 2 Beaufort Place in Clifton, was responsible for all fourteen of these Gloucestershire based units as well as the various cadets, finding suitable premises for them as well as providing equipment and clothing; arranging training facilities and annual camps; recruiting; and meeting running costs. In 1909, the Gloucestershire CTA, mainly composed of prominent businessmen under the chairmanship of the lord lieutenant, queried the amount of time it took to get answers from what they saw as an inefficient and dilatory War Office. Among the many new drill halls, commissioned across the country by the CTAs, was Old Market Street in Bristol, which opened in 1915. Additional to those Gloucestershire units spread across the county, a company of the 1st Battalion,

Herefordshire Regiment, an all-TF unit, was based at Ruardean with a drill station at Littledean, with a further company of the 4th Battalion, Oxfordshire and Buckinghamshire Light Infantry at Stow-on-the-Wold.

In 1909, the Bristol Division of the Royal Naval Volunteer Reserve (RNVR) moved from their drill ship to 37 Jamaica Street, built as a carriage works in 1905. Royal Naval Reserve (RNR) and RNVR personnel would henceforth put in time at sea rather on the drill ships which were mainly disposed of with HMS *Daedalus* being sold for scrap in 1911.

The Beginnings of Military Flight in Gloucestershire and Bristol

Promoted by Sir George White, the entrepreneur, aeroplanes began flying from Filton in 1907 as a means of coaxing passengers out to his new tramway terminus. Using the tramway shed, aircraft construction commenced in 1910 with the Bristol Boxkite, built by White's British & Colonial Aeroplane Company. It was hoped to employ aircraft for reconnaissance in the annual military manoeuvres of 1910 but neither aircraft nor pilots were available via official channels. Sir George stepped in, offering a Boxkite to be flown by his own pilot, Bertram Dickson, a former artilleryman. Despite early setbacks, both technical and meteorological, Dickson was able to demonstrate the usefulness of airborne reconnaissance by locating Blue forces and notifying the Red HQ of their position. Despite this success, the army declined to show any real interest in aviation for a further three years, although Haldane as minister for war attempted to expand the RE Air Battalion by ordering four Boxkites in early 1911. Within a very short time, the Boxkite was regarded as obsolescent and Sir George engaged a Frenchman, Pierre Prier whose first attempt at a monoplane was not a success and Bristol's temporarily dropped out of the running for the Air Battalion's new aircraft. In 1912, Prier's machine flew again but was grounded after repeated engine failures and its fate was sealed by a blanket ban on monoplanes. However, an airfield had been established at Filton with permanent iron hangars and manufacturing facilities soon to come into their own.

5

Gloucestershire and Bristol in the First World War 1914–18

The outbreak of war in August 1914 found Britain's forces largely unprepared and under-manned. The regular army quickly mobilised its reserves to bring its battalions up to strength prior to a deployment in France and Flanders, while the territorial units assembled in their regional divisions to take up their assigned home defence stations. As soon as Kitchener was appointed war minister, he called for the immediate recruitment of 100,000 men for his New Armies, and the TF stepped up recruitment. Offices opened in Bristol's Colston Hall on 13 August, with others in drill halls and town halls across the county. All military establishments moved onto an immediate war footing and on 14 August, the vast Shirehampton Remount Depot opened. As well as remaining the Glosters' depot, Horfield Barracks became No. 5 (Southern) Cavalry Depot, then No. 5 (Hussars) Cavalry Depot for the 3rd, 7th, 15th, and 19th Hussars, providing fresh recruits to the 7th Hussars in Mesopotamia and the other three regiments in France for the duration. In moves to counter what was perceived as immoral behaviour in towns newly flooded with military personnel, patrols were set up by women's organisations along with alternative entertainments to drinking, gambling and whoring. In Bristol in 1915, the inaugural Training School for Women Patrols and Police was set up, influencing developments elsewhere in the country.

The nearest the Owen Committee, reporting on coast defence needs in 1905, had got to the Severn were Cardiff and Barry, judging them to be low priority. The Severn defences were stripped, and submarine mining was handed to the Admiralty which promptly scrapped the whole system. By 1914, however, the need to revive the Severn defences had been realised as protection for the volume of mercantile shipping into Bristol, Avonmouth, and Gloucester. By early 1916, a pair of naval 4.7-inch QF guns had been installed on Avonmouth's South Pier, and two naval 12-pounder QF guns in the old Portishead Point Battery which commands the deep-water shipping channel. These guns were a defence against surreptitious raiding by U-boats and opportunistic strikes by torpedo boats. Using a combination of fixed-beam searchlights and pre-ranged guns, well-drilled gun-crews could be confident of picking off intruders.

The Deployment of Local Units

In August 1914, troops rushed to join their parent formations. The 1st RGH, reorganised into three squadrons to mirror the regular cavalry, assembled in a camp on Newbury Racecourse before a move to East Anglia with the South Midland Mounted Brigade. They then sailed for Alexandria whence they went to Gallipoli as dismounted troops, returning to Egypt to be reunited with their horses, reverting to the cavalry role for the rest of the war, fighting their way up through Palestine and into Syria. The 2/1st South Midland Mounted Brigade, which included the 2/1st RGH, formed at Cirencester Park in April 1915 and was sent to Norfolk where they were converted into a cyclist brigade, a move which provoked dismay. The 3/1st RGH formed later in 1915 and spent the war as a Reserve Cavalry Regiment in Tidworth, providing drafts to the 1st and the 2/1st RGH.

The Glosters' 1st Battalion, stationed at Bordon in Hampshire, joined up with the 1st Infantry Division to sail from Southampton to Le Havre on 13 August. Leaving a cadre at Horfield Barracks to man the depot and receive new recruits, the Glosters' 3rd (Reserve) Battalion left Bristol to join the Thames and Medway Garrison where it would remain throughout the war. The Glosters' 2nd Battalion sailed from China, landing at Southampton in November, to arrive in France in December. The Glosters' 1st Battalion spent the entire war on the Western Front, but the 2nd Battalion went on to Salonika in 1916.

When war broke out, the 4th and 6th Battalions had been at Minehead for their summer camp. They hurried back to their drill halls and along with the 5th Battalion were sent to Essex as anti-invasion troops in the 48th (South Midland) Division as part of Central Force's Third Army with its HQ at Luton Hoo in Bedfordshire. However, to make up for the disastrous losses incurred during the Retreat from Mons, they would be hurriedly deployed to France in May 1915. All three Glosters TF battalions would also spend time in northern Italy during the course of the war. The three second-line 2/4th, 2/5th, and 2/6th TF units were raised to join the 61st (2nd South Midland) Division, deployed across Bedfordshire and Hertfordshire, before themselves being despatched to France in May 1916, where they remained. The three third-line battalions combined as a reserve unit, performing home defence and training functions all over the country including a spell based in Cheltenham.

Some eight service battalions of Kitchener's New Army were badged as Glosters during the war, and two of them, the 12th and 14th Battalions were raised in Bristol by the Citizens' Recruiting Committee in, respectively, August 1914 and April 1915. The 12th was a true 'Pals' battalion, and the 14th was a Bantams battalion for men under the regulation height of 5 feet 3 inches (1.6 metres) and both these battalions were accepted by the War Office in June 1916. The 13th Battalion was recruited mainly from miners in the Forest of Dean and South Wales and was accepted by the War Office as a Pioneer unit. These three battalions, along with the conventionally recruited 8th and 10th Battalions, spent their war in France. The 7th Battalion fought at Gallipoli where it suffered over 800 casualties including its entire leadership. After reforming, it remained to fight in the Middle East campaigns. The 9th Battalion left France for Salonika in late 1915 where it stayed before returning to France in the summer of 1918. The 18th Battalion was assembled late in the war and also fought in France. Additionally, the 11th, 15th, and 16th Battalions served in Britain as Reserve Battalions training newly conscripted recruits from 1916 onwards. Home Service TF troops who, mainly for reasons of age and fitness, had elected to serve at home, comprised the 17th (Provisional) Battalion, based in Essex on home defence duties, brigaded with other such units. The Glosters thus maintained a total of twenty-four battalions at different times over the course of the war, of whom seventeen served overseas. Over 8,000 men lost their lives serving with the regiment and numerous medals were won including four VCs.

The gunners of the 1st South Midland Brigade RFA (TF) from Clifton and Gloucester gathered at the Whiteladies Artillery Ground to commence intensive training in England prior to leaving for France with the 48th Division in 1915, equipped with their old 15-pounder guns. In 1916, they were reorganised into three batteries of 18-pounder guns and a new fourth battery with 4.5-inch howitzers, and in 1918, they were redeployed to Italy. Their second-line formation, raised in Bristol in 1914, remained in England until 1916 when it, too, was sent to France with the 61st Division. It spent much of its time in a passive form of trench warfare, equipped with French 90-mm field guns. The Citizens' Recruiting Committee also raised two heavy batteries of the RGA, the 127th and 129th, each almost 200-strong, with many of those recruited to the 129th coming from the Gloucestershire and Bristol police forces. Following extensive training at White City (see below), both Bristol batteries were deployed to France in the spring of 1916. The first- and second-line ASC and engineer units accompanied their respective divisions to France, the sappers of the second-line company contributing to tunnelling and mining operations associated with the Battle of the Somme. The 3rd (South Midland) Field Ambulance, RAMC, TF, served with the 48th Division in France and Italy and its second-line unit served in France. To meet the demand for organised labour under military control, labour companies were formed, enlisting some who had served and been wounded or who were considered no longer fit for military service. In May 1917, these included No. 313 Company, recruited in Bristol and No. 314 Company in Shirehampton. Later that year, Nos 440, 599, and 648 Agricultural Companies were raised in Bristol.

Early in the war, suitable accommodation was scarce, and the newly opened Ashton Gate Exhibition Centre, known as the White City, the site now occupied by Bristol City Football Club, was purchased by the War Department for use as drill hall, canteens, barracks, mess rooms, and rifle range, all occupying the replica castle, Shakespearean-age houses, and funfair. This became home to the 900 recruits to the Glosters' 12th (Service) Battalion who remained there until they joined 95th Brigade, 32nd Division, in June 1915, proceeding to Salisbury Plain and thence to France in November 1915. They were followed by the RGA gunners of 127 and 129 Heavy Batteries. The 9th and 11th Battalions were fortunate to spend their first winter in billets in Cheltenham. The Park Street drill hall had been absorbed into the new Wills Memorial building, built for the university in 1913, and was replaced by the new drill hall in Old Market Street which opened in June 1915, paid for by the Wills family.

The Senior OTC at Bristol University was only fifty-five-strong in 1914, but the university was selected as one of only a dozen Officer Cadet Training Battalions, using the university buildings and facilities. Some 700 officers were trained there, of whom 105 lost their lives. Those schools with Junior OTC units had also prepared themselves for war. Clifton College gave military training a higher profile than many schools and had an inter-house competition, called the Canada Cup, for promoting military skills. It offered its premises and its rifle range for the use of those training Bristol's new units. By 1918, some 3,000 former pupils had served in the armed forces and almost 600 had lost their lives. Clifton College's OTC, an engineer corps and, as such, one of only a handful of such units in the

Bristol, the Old Market Street drill hall now converted into apartments.

Bristol, the Whiteladies artillery drill hall still in use as an Army Reserve Centre.

country, prepared many cadets for commissions in the REs, as well as in other branches of the services, practising bridge-building and launching pontoons on the ponds of Bristol Zoo, opened in 1836. Many of the hundreds of officers needed by all the new battalions came from these various cadet units: the OTC, the TF cadets, the Boys' Brigade, and the Church Lads' Brigade.

The Shirehampton Remount Depot was built to receive the large number of horses and mules being imported from the USA, Canada, Spain, and South America, through Avonmouth Docks, to meet the army's needs. Each of its thirty-five paddocks could accommodate 100 horses and quarantine them effectively prior to onward despatch, and its veterinary department was equipped to house 200 sick horses. The depot was staffed by up to 1,500 men of the Army Service Corps, the Army Remount Service, men and women of the Forage Corps, and members of the Women's Auxiliary Army Corps providing administrative and domestic support, all accommodated in corrugated-iron barrack huts. The site also contained forage barns; offices; stores; canteens and messes; farriers' forges and shoeing sheds. The railway line into Avonmouth Docks ran between the depot and the north bank of the Avon. The site extended from the big roundabout by Avonmouth Industrial Estate down to Springfield Avenue and Station Road. Over a third of a million equines passed through the depot during the war and Manor Farm at Quedgeley became a convalescent depot for shell-shocked horses with a purpose-built stable block added to the existing farm buildings.

The Volunteer Training Corps

In 1914, the perceived threat of a German invasion, exacerbated by scare stories which saw German nationals, resident in Britain, performing acts of sabotage, caused a frenzy of identifying and guarding 'Vulnerable Points'. Boy Scouts guarded the Clifton Suspension Bridge and took over some coastguard duties such as manning the Walton Bay signal station at Clevedon, looking out for U-boats and reporting to the coast battery at Portishead. The 5th Battalion, Loyal North Lancashire Regiment, was stationed at Chipping Sodbury, centred on Hartley House and tasked with guarding the railway line into Avonmouth Docks. Sadly, owing to an unfamiliarity with the environment, one of their number was killed by an express train in the first week, and a second soldier a week later. Some elderly reservists served as full-time soldiers in the Protection Companies which would later become the National Defence Corps. However, seeing the need for local men over the age of enlistment or in essential occupations to be seen to be involved in the war, and spotting the tasks to be performed, across the country, groups of volunteers offered their services to the military to release regular troops to the frontline and to supplement those on home service. Initially, these local groups, recruited from rifle clubs, neighbourhoods, social networks, or occupational groups, acted independent of authority but the War Office reluctantly accepted a national committee under military oversight to give them some respectability. In Bristol, the Coliseum Volunteer Corps, the Rugby Volunteer Corps, and the Bristol Savages VTC formed, along with similar groups in Berkeley, Gloucester, Cheltenham, Thornbury, and many other places across the county, under the generic title of Volunteer Training Corps (VTC). Under the direction of old soldiers, units trained on evenings and weekends; Gloucester units were drilled by the local police using dummy rifles; and Bristol University ran evening classes in military matters for VTC members. By June 1915, all these small individual groups were brought together into county Volunteer Regiments (VR). June 1915 saw the formation of the Gloucestershire VR, with Earl Beauchamp as president, Colonel Cecil Davis as regimental commandant, and Theodore Hannam of Queen Street, Gloucester, as correspondent. The VR established its HQ and drill ground at Kingsholm, with Forest of Dean, Gloucester, and Cheltenham battalions. Attempting to attain a battalion size of around 1,000 men, the Cheltenham and Gloucester battalions soon merged. The County of Bristol VR shared an administrative HQ with the National Reserve on College Green with an Adjutant's HQ in Queen Square. The operational RHQ was at the County Rugby Ground and then, from 1917, at the St Michaels Hill drill hall. In 1916, the War Office finally accepted the VTC as a military force in principle and sanctioned uniforms and weapons. Now to be known as the Volunteer Force (VF) it was reorganised with the volunteer battalions badged to the Glosters. Only historic counties with lords lieutenant were permitted, so the three Bristol battalions were required to 'assimilate' into the county force, becoming the 2nd, 5th, and 6th Battalions of the Gloucestershire VR, alongside the

1st (Forest of Dean), the 3rd (Cheltenham & Gloucester), and the 4th (Berkeley Vale & Stroud) Battalions. In Bristol, the Brecknell, Munro & Rogers Cadet Engineer Field Company and Kingswood Training School Cadet Corps were affiliated to the 2nd Battalion, along with the Lord Mayor's Own Cadet Corps. Gloucester's Wagon Works Cadet Corps, and the St Paul's Cheltenham Cadet Corps were both affiliated to the 3rd Battalion, while the Stroud Cadet Corps was affiliated to the 4th Battalion. The Brimscombe Polytechnic Cadet Corps, with an entry in the Army List and based in their school-building in the old canal warehouse which reopened in 1911, was most probably also affiliated to the 4th Battalion. Corps were based in whatever premises were made available to them. Thornbury VTC which became 'C' Company, 4th (Berkeley Vale) Battalion, had its HQ in Thornbury Castle and held training sessions in Thornbury's Castle Hotel. The Frenchay unit met in the village hall and space was set aside in Sudeley Castle for the local Winchcombe corps to drill. In the early days, corps determined their own activities, training schedules and drills. The Coliseum Volunteers laid a pathway at Southmead Hospital anticipating increased use as it came under army control, while the Brecknell cadets set up a workshop to manufacture munitions. Units drilled locally and met up for field days and camps, and in August 1916, the Bristol battalions invited corps from Northamptonshire to share a camp with them. Other cross-corps links were forged by shooting competitions. One activity carried out regularly from May 1915 by Gloucester Volunteers was patrolling the city streets from early evening until midnight. By 1916, this had been extended to nightly patrols across the whole city as an early warning against Zeppelin raids. The Berkeley Vale Battalion had been given the task of patrolling the railway from the earliest days of the war. Bristol Volunteers put in shifts at the docks unloading timber. Most of this activity was financed by members' subscriptions, patronage, and local authority funding but Cirencester was criticised for being the only local council unwilling to contribute. Once the War Office had integrated the VF into the nation's anti-invasion plans, an event celebrated by a parade at Gloucester Cathedral, specific tasks of a military nature were allocated, supported by government funds. Bristol's three battalions, along with the two Glosters battalions based in Gloucester and Stroud, were part of a force of twenty-seven battalions securing the railway lines in the south-west which constituted No. 1 Lines of Communication with its HQ at London's Waterloo Station. The third Glosters battalion, concentrated at Dursley, was under orders in an emergency mobilisation to proceed to the Ripon Reserve Centre in North Yorkshire. A dress rehearsal mobilisation was carried out in March 1916, shortly followed by an unannounced trial run. In the summer of 1918, the country had been stripped of troops to counter the German 'Michael' offensive and the VF was asked to provide Special Service Companies to take over coast defence duties and so release TF cyclist troops for a more mobile role. The company provided from the Gloucestershire VR, composed of men who could be spared from their day jobs for three months, was assigned to East Anglia.

The Maritime War

Although the Severn ports accounted for significant proportions of Britain's imports and exports via Avonmouth Docks, Bristol's Floating Harbour, and Lydney Harbour on the north bank, there was only a minimal naval presence. Bristol's RNR Division retained its HQ at 37 Jamaica Street throughout the war and was the HQ of the Coastguard and Reserves Western District which stretched for several hundred miles from Yealm, east of Plymouth, to Torrs Point, south of Kirkcudbright. Coastguards belonged to the RNR and were mobilised for service at sea, their place being taken by wives and Boy Scouts. The RNVR had an excess of stokers who, as a consequence of so many naval vessels converting to oil, had become redundant. Many from this surplus pool were therefore mobilised as infantry in the Royal Naval Division, fighting at Antwerp, Gallipoli, and the Western Front. Those naval personnel who remained were served by HMS *President III* as their accounting base from recruitment through to demobilisation. It also administered the Defensively Armed Merchant Ships (DAMS) training programme up until 1918, and the minesweeping flotillas in the Severn. Although the Severn never constituted a naval base as such, precautions were nevertheless put in place to defend such important trading ports. Merchant ships, convoyed into the Severn estuary, used shipping lanes which needed to be kept clear of mines. P & A Campbell's White Funnel Fleet's thirteen pleasure steamers, with their crews, were all requisitioned for minesweeping operations and two of them, the *Brighton Queen* and the *Lady Ismay* were lost, but without loss of life. One even sailed as far as Alexandria and ferried a surprised party of RGH to Gallipoli. Nine Bristol-registered trawlers were also requisitioned by the Admiralty, one of which was lost to a mine. HMS *Essex*, a depot ship, is listed as being based at Avonmouth for the first six months of 1917. Two 11,000-ton liners which had sailed on regular routes across the Atlantic between Avonmouth and Canada were commandeered as troopships and renamed as HMTs *Royal Edward* and *Royal George*. They brought Canadian troops into Avonmouth *en route* for the Western Front. In July 1915, the *Royal Edward*, having served briefly as an internment ship off Southend-on-Sea, left Avonmouth with over 1,300 reinforcements for the Dardanelles campaign. She was torpedoed off Greece by *UB-14* with the loss of nearly 900 lives. Closer to home, February 1917 saw a renewal of the German policy of unrestricted warfare which resulted in the loss to submarine attack of much shipping off the South Wales coast. Ships in and out of Bristol and Lydney are recorded as being among the dozen sunk by *UC-65* in one week. A total of seventeen colliers in Admiralty service are listed as being lost in the St George's or Bristol Channels, to either mines or torpedoes. Despite such events, Avonmouth and Portishead Docks and Bristol's Floating Harbour continued to be seen as a safe distance from the action and saw the arrival of vast quantities of imports, especially food, which made use of Avonmouth's cold storage facilities with smaller vessels carrying vital food supplies right into the heart of Bristol. As well as the horses we have already met, troops from the empire and war materiel in great quantities also arrived in the Severn. Oilers brought large volumes of petrol which was then repackaged in 4-gallon (18-litre) tins and distributed from Portishead Docks to Britain's increasingly mechanised armies. Although Gloucester's docks had lost much of their trade by the early years of the century, some commodities such as timber maintained some importance and managed alterations to the river flow aided riverine traffic as far upstream as Tewkesbury.

The War in the Air

Throughout the war, Gloucestershire's contribution to military aviation primarily consisted of aeronautical development, aircraft manufacture and repair, and flying training. By 1914, Filton was well established as an aeronautical centre, manufacturing aircraft and training pilots. Its first squadrons formed in early 1916 flying B.E.2s for training prior to deployment over the Western Front. However, the airfield's primary function would be as an Aircraft Acceptance Park (AAP) and this emphasis was reflected in September 1916 when No. 21 Training Wing RFC was transferred away to Cirencester Park. New RFC pilots received two months' general military training before moving up to a School of Military Aeronautics (SMA) where they were instructed in the theory of flight. They were then moved on to a training squadron based on an airfield. Towards the end of the war, there appears to have been a concerted effort to concentrate these various threads. In January 1918, No. 9 SMA moved into Cheltenham's Marlborough House from Coley Park Reading, and just before the war ended, No. 6 SMA moved from Denham (Buckinghamshire) to 10 Elmdale Road, Clifton. Rendcomb had opened in June 1916 as a training station, soon to become No. 38 Training School (TS). The next year, the Australian Flying Corps (AFC) established their HQ in Tetbury and opened two training stations, Minchinhampton in late 1917 and Leighterton in January 1918. Lasborough Park Mansion housed Leighterton's officers' mess and junior officers were billeted at Lasborough Manor House. Rendcomb moved up a notch to become No. 5 Training Depot Station (TDS) in August 1918 expecting to have to deal with an influx of US pilots converting to the larger four-engined bombers such as the Handley Page 0/100 and improved 0/400, biplanes with a range of 700 miles (1,100 km), a bombload of 2,000 lb (907 kg) and a top speed of 97 mph (157 kph). With the end of the war, Rendcomb was downgraded to a TS. The aircraft flown by these training squadrons were mainly the Avro 504K; the Sopwith Pup, Camel, and small numbers of the Snipe; the F.E.2B, S.E.5 and B.E.2C; and the DH.5 and DH.9. With the RFC and the RNAS competing with each other to commission new models, often with little in the way of evaluation carried out, and an arms race with the German air force, each side achieving temporary ascendancy in turn, a mechanism for receiving new aircraft and putting them through their paces became necessary. Aircraft Acceptance Parks (AAP) were built at Filton and Brockworth. No. 5 AAP at Filton received the output of the British & Colonial Aeroplane Co. known as Bristol; of Westland & Whitehead building DH9 and DH9A aircraft; and of the Gloucestershire Aeroplane Company which, as HH Martyn of Cheltenham, built F.2Bs. The park was served by a (2-foot/61-cm) narrow gauge railway with two steam locomotives. Brockworth began as Martyn's aircraft testing airfield in Hucclecote but was chosen as the site of a new AAP in late 1917. As well as receiving new aircraft, the RFC determined to repair as many of those aircraft damaged in accidents as possible. The Western Aircraft Repair Depot (ARD) was set up at Yate early in 1917 and later in that year it became No. 3 (Western) ARD and by the spring of 1918 was employing several hundred personnel of the Women's Royal Air Force (WRAF) and aircraft repair squadrons of the US forces. Avonmouth Docks was the embarkation port for at least one RFC Aircraft Park on its journey to France.

All these establishments needed large numbers of capacious buildings in order to shelter the hundreds of aircraft on inventory. The Flight Shed of 1915 was a front opening gabled rectangular shed which came in lengths of 210 or 140 feet (65 or 43 metres) and a span of 65 feet (20 metres). The slightly larger 1917 General Service (GS) Aeroplane Shed had a bowstring roof supported on timber 'Belfast' trusses, spanning the entire 65-foot (20-metre) width, enabling aircraft to be shifted around without the encumbrance of supporting pillars. It measured 170 by 80 feet (52 by 24.5 metres) or 170 by 100 feet (52 by 31 metres) and came in triple or single versions. As aircraft got bigger, so also did the need for multiple hangars and the storage sheds for the Handley Page bombers measured 200 by 60 feet (61.5 by 18.5 metres) being erected in triple units 180 feet (55 metres) across. Rendcomb TS had six flight sheds; Brockworth AAP had five coupled GS sheds and seven triple-unit Handley Page storage sheds in 1917; Filton had five flight sheds and thirteen GS aeroplane sheds as four triple-units and one single; and Leighterton had four coupled GS sheds and a single aircraft repair shed (ARS) along with two aeroplane stores. At Minchinhampton, four GS aeroplane sheds were built along with ten canvas Bessoneau hangars. Yate had a more motley collection of buildings including erection sheds and engine workshops, all covering 40 acres (16 ha). All these sites had numerous other buildings for technical, instructional and domestic functions. For the most part, huts of different types, many of them brick-built but other materials such as timber, plasterboard and corrugated iron were used, and tents were not unknown. There are some impressive remains of some of these structures. At Filton, one of the triple GS aeroplane sheds stands, along with the single GS shed and one of the two flight sheds. Some of the flight sheds and the brick-built station workshops and flight office at Rendcomb have been cleverly restored and continue to fulfil aviation functions more than 100 years on. In Yate, Poole Court survives from the ARD days.

Rendcombe airfield, the restored First World War station workshops.

Filton airfield, a First World War flight shed.

Filton airfield, the triple First World War GS hangar.

The Aircraft Industry

In 1914, Sir George White's British & Colonial Aeroplane Company, known as Bristol, at Filton was soon producing the RFC's preferred B.E.2, two-seater biplane, a total of 1,100 in all, but also developed the Bristol Scout for service with both the RFC and the RNAS. Fortunately, the opportunity fell to Bristol to design and produce the next generation of two-seater fighter. This was the F.2B, better known, with its newly developed Rolls-Royce Falcon engine as the Bristol Fighter or 'Brisfit' and adopting the same technology as Fokker to synchronise the firing of the machine gun through the propeller. By the end of the war, over 3,000 Bristol Fighters had been built at Filton and at the Brislington Tramways factory, while many more were built under licence elsewhere. Rolls-Royce Falcon engines and their Hawk engines for naval airships, as well as Renault engines for trainer aircraft, were all built at the Brazil Straker works in Fishponds. In 1918, the Bristol Triplane bomber was nearing production but was too late to go into service. It could fly at 125 mph (200 kmh), carrying a bombload of 2,700 lb (1,227 kg), over a range of 500 miles (800 km). Thankfully, the Triplane would only ever have to fly as a fourteen-seat passenger aircraft after the war.

The Gloucestershire (later Gloster) Aircraft Company, trading as HH Martyn, rented space at Sunningend, Cheltenham in 1917 to fulfil contracts for AIRCO which part-owned the company, building 150 Airco DH.6 biplane trainers by the end of 1917. By 1918, Gloster was producing forty-five Bristol Fighters a week. Aircraft fuselages, minus their wings, were transported to the Brockworth AAP by road. HH Martyn used Sunningend Factory and leased Cheltenham Winter Gardens for aircraft production. During the war, in all, nearly 5,000 Bristol Fighters were built. Parnall & Sons, using Filton airfield for flight testing, assembled Short Seaplanes, Scout Seaplanes and Bombers, and Avro trainers at their Park Row works, the Coliseum roller skating rink in the centre of Bristol, building nearly 1,000 aircraft by the war's end. In Dursley, Mikael Pedersen, a Danish bicycle manufacturer, designed a magneto for the aircraft industry, replacing the German Bosch instrument which had hitherto met the industry's needs.

War Production

Munitions: Shells and Bombs

No. 5 National Filling Factory was set up at Quedgeley in 1915 with offices in Apsley House, Gloucester and employing mainly female staff recruited from domestic service. The site was served by a railway line and had an internal narrow-gauge layout. Filled shells were stored in twenty magazines and black powder and cordite were held in dedicated magazines. By March 1917, there were over 6,000 employees, 90 per cent of whom were women. The factory is recorded as manufacturing over 10.5 million 18-pounder and 4.5-inch howitzer shells, 17,000 shrapnel shells for 60-pounder guns, 7 million filled cartridges for guns from 2.75-inch to 9.2-inch, and 3 million fuses. The buildings were demolished 1924–6 but one may remain at 330 Bristol Road, Severn Rewinds Ltd, where a Belfast Truss shed, standing opposite Byars Road (SO818162) is reputed to have come from the filling factory. One account has the shed arriving by barge so it may rather have come from the Slimbridge magazine. A second such shed off Clifton Road (SO828176) appears now to have gone. The Castle Iron Works, on Victoria Road, Bristol became a manufactory for 18-pounder shell cases and then as a shell-filling plant turning out 160,000 filled shells and 300 tons of cartridges per week by the end of the war. Daniels of Stroud and the Gloucester Wagon Works also manufactured shells, and Lister's of Dursley made shell cases and fuses. Firms producing glycerine, even as a by-product had their product requisitioned by the explosives industry to the extent that their main production was skewed. Even firms such as WD & HO Wills (tobacco) and JS Fry (chocolate) converted parts of their production lines to making shell cases. Any company with spare capacity or empty premises turned to producing munitions. Brecknell, Munro and Rogers, who had raised a cadet company from among their younger employees, converted an empty chapel they had previously used as a store, into an engineering workshop turning out shell cases. Capitalising on the natural materials to be gathered in the Forest of Dean, HM Acetone Factory stood on Speech House and Monmouth Roads in Coleford. By 1916, Crane & Company at Warmley, firework manufacturers, were filling between 50,000 and 100,000 Mills Bombs with ammonal. The largely female workforce styled themselves the 'Warmley Grenadiers'.

Chittening Trading Estate, Avonmouth, was the site chosen by the Ministry of Munitions and the Trench Warfare Department for a mustard gas factory on St Andrews Road (later to become Britannia Zinc). This secret site, National Filling Factory No. 23, was run by Nobel's Explosive Company, one part producing the gas, known as HS, and another filling 6-inch shells and bombs. Holland & Cubitt's Grillo Oleum Plant Absorption House processed sulphur for explosives. Needing to expand production, the redundant picric acid plant in Avonmouth was taken over for an increase in production. Sickness and injury, directly caused by unregulated contact with poisonous and corrosive substances, within the predominantly female workforce, caused severe disruption to production and lasting physical damage to the workers over 1,200 of whom suffered severe illness from exposure to toxic substances. Storage of the output of the munitions factories posed a logistical problem, and it soon became necessary to establish new magazines. Munitions were stored in Monk Meadow beside the Sharpness Canal in Gloucester, but further expansion was needed and one of two new sites chosen in 1915 was Slimbridge. The site, officially HM No. 23 Magazine, Gloucester, was acquired through DORA and lay west of The Moors between Tumpy Green and the Sharpness Canal. It was served by branch lines which left the main Midland Railway line at Gossington Bridge. Sixteen timber magazine buildings, raised on concrete pillars to railway platform height, each held 200 tons of cordite propellant. Three magazine buildings held 1,100 tons of gunpowder. Near Old Hurst Farm were the locomotive sheds and workers' barracks for the eighty staff. In August 1917, a new magazine with a capacity of 2,000 tons was constructed just north of Tumpy Green Farm. The whole complex was surrounded by a fence and guarded by a company of the Royal Defence Corps. At the end of the war, the cordite was destroyed by fire and the buildings were auctioned. By the mid-1920s, the site had returned to agriculture.

Other Elements of the War Effort

As a response to the large volume of merchant shipping being sunk by U-boats, National Shipyards were established by the government. The first was an existing shipyard, built in 1916 at Chepstow at the mouth of the Wye but taken over by the government and expanded to eight berths. The second was to be at Beachley and would be larger with eighteen berths, and the third, with eight berths, would be where Portbury Docks now stand. Despite the expenditure of £2 million at Beachley alone, taking place late in the war, and being subject to all sorts of delays, only one ship was built at Chepstow, and none completed at the other two yards. Jetties and railways at Beachley, built by REs and German POWs, are recorded on the 1921 OS map, along with housing for workers at Sedbury's Pennsylvania estate, still largely recognisable. At Hempstead from 1917, 350 men were engaged in constructing six concrete barges. Mawdsley's of Dursley were producing dynamos and motors. The value of motorcycles in military operations had been demonstrated in 1910 when they were used for reconnaissance and despatches. Already producing generators for the War Department, and following successful trials, the Douglas Company of Kingswood was chosen to produce motorcycles for the services. Large numbers of stationary engines used for everything from powering wirelesses to filling hot-air balloons were supplied, often in short order. As well as these, 25,000 motorcycles, supplied to the War Department, were used in all theatres of the war including Russia and the wilds of Central Asia and Africa. John Lysaght, previously a supplier of fencing and ironwork for farms, switched to barbed wire and steel sheeting for barrack huts and aircraft hangars. Tracked vehicles were a British invention which had failed to attract the attention of the military and the technology had migrated to the USA. By 1915, their usefulness for towing heavy artillery in the Flanders mud had finally been recognised and tractors were imported through Avonmouth Docks. A large park was established where they could be prepared for onward delivery to the army. Lydney Harbour was used for exporting coal and iron ore from the Forest of Dean and coasters using the harbour became the targets for U-boat attacks. To meet the vastly increased need for quickly erected buildings, National Factory No. 212 was opened in Yate by Robert MacAlpine and Sons, manufacturing concrete slabs and reinforced concrete joists and girders. Joseph Griggs & Co. on Bristol Road, Gloucester, ran a sawmill under government direction. Britton & Sons in Kingswood received contracts to make army boots, and Wathen Gardiner & Company, a clothing firm of Staple Hill, Bristol, made military uniforms up until the night their new factory was burnt down in January 1917.

Alongside all this industry contributing directly to the war effort, everyday life had to go on: the people had to be fed and clothed; public transport had to run; and the sick had to be tended. Allotments, poultry-keeping, and foraging activities helped to supplement farm-grown produce and went towards compensating for the lost imports sunk by the U-boats. Members of the Women's Land Army, schoolchildren, POWs, and troops all helped on the farms all the year round but particularly at harvest times.

Military Hospitals

The Bristol Royal Infirmary at Kingsdown, was the designated home of the 2nd (Southern) General Hospital, RAMC (TF) which quickly absorbed the Southmead Workhouse Infirmary increasing its capacity from 260 beds to 1,300. Bishop's Knoll House, at Stoke Bishop, also formed part of this 2nd General group. Its owner, who had spent many years in Australia, dedicated the mansion as a 100-bed war hospital serving ANZAC casualties who were brought in via Temple Meads station. Also under the auspices of the 2nd Southern General was the Almondsbury Subsidiary Hospital, known as the Institute before the war and as the Cottage Hospital afterwards. At Fishponds, the Glenside City and County Asylum became the Beaufort War Hospital with an annexe at Tockington. Auxiliary military hospitals were organised by the British Red Cross and the St John' Ambulance Association; by other organisations under the Voluntary Aid Detachment (VAD) umbrella; and by individuals opening their houses as convalescent homes or hospitals. The Cheltenham Racecourse Red Cross Hospital was in use for wounded Belgian soldiers by November 1914, filling its 103 beds in the Grandstand. By 1918, huts had been added, doubling its capacity. Municipal buildings serving as hospitals included Bingham Hall in Cirencester, the Assembly Rooms in Winchcombe, and Tewkesbury's Watson Memorial Hall. Some sites needed to be upgraded as the war produced ever more casualties, the fifty-bed Hillfield House in Gloucester transferring to the 100-bed Bishop's Palace, and Lydney Town Hall expanding into the adjacent secondary school. Private mansions given over for medical use included Boddington Manor, Leckhampton Court, and Tockington Manor. Very often, as was the case at Bishop's Knoll and Cleve Hill, the resident family took on the responsibility for financing and managing the establishment. Schools also served as Kingswood School in Bristol converted to an auxiliary hospital in 1916; Suffolk Hall in Cheltenham became a sixty-five-bed hospital; and the Red Maids' School at Westbury-on-Trym accommodated 200 beds. Some hospitals such as Ashton Court and Forthampton Court were exclusively for officers. As well as the larger hospitals, smaller establishments such as the twenty-bed Ghyll House at St Briavels, the twenty-eight-bed Red Cross hospital in Horton, the thirty-bed Trinity Parish Rooms in Stroud or the thirty-bed Dumbleton village hall, staffed by No. 54 VAD, provided valuable supplements to the larger hospitals. Some such as the fifty-bed Copse Hill House at Bourton-on-the-Water provided for specifically convalescent patients, whilst others became specialist centres offering higher levels of care. Portcullis House at Badminton became an annexe of the Roehampton Hospital in west London, caring for limbless soldiers awaiting treatment at the parent hospital. One of the duties of the various volunteer motoring organisations was the transport of wounded men from railhead to hospital. Where the hospital lay close to the railway station, as did Portcullis House in Badminton, this task would prove less stressful for driver and patient alike.

Cirencester, Bingham Hall, plaque to commemorate its use as a First World War military hospital.

Prisoner of War (POW) Camps

During the First World War, POW camps seldom resembled the camps of the Second World War with wire fences and watchtowers. One camp that did fit the stereotype was at Yate. Here, to the east of Westerleigh Road, a camp was originally built for civilian internees, aliens from enemy countries, who were then moved to the Isle of Man. It reopened for combatant POWs in March 1917 consisting of twenty wooden huts, each measuring 80 by 30 feet (25 by 9 metres) which accommodated a total of 500 POWs. Along with an ablutions block, a hospital, a canteen, and other structures, they were set in a fenced enclosure measuring 100 by 150 yards and surfaced with cinders. Some of those German POWs were employed in the construction of Yate aerodrome in 1917 as putting POWs to work appears to have been the norm. In September 1917, the inhabitants of Beachley were moved out of their homes to clear the space for a new National Shipyard. A railway was laid through Sedbury and a POW camp was built. By the next March, 2,000 POWs had been installed in fifty-four timber huts, their job being to cast concrete blocks for the shipyard workers' housing. At Tytherington, fifty Austrian POWs worked in the stone quarries, and at Flax Bourton fifty POWs, engaged in agricultural work in 1918, were housed in the schoolhouse of Bedminster Workhouse. Some camps were only temporary as at Henbury where four large and six smaller huts were put up for a short period in 1917, to house 400 POWs working on construction and road building but closed within a short time of opening. Penpole Camp, Shirehampton, was a tented camp for 100 POWs employed in small working parties on roads and buildings but only operated for four months. Marlwood Grange, at Thornbury, housed forty-two POWs working on the land in 1918. There were further camps at Cheltenham, Churchdown and Toddington.

6

Gloucestershire and Bristol during the Interwar Years 1919–39

The Sea Mills housing estate, built as a garden suburb in 1919, was the first of many such projects begun by municipalities after the war. As 'Homes for Heroes', along with cenotaphs, memorial halls, village war memorials, and edifices such as Clifton College's Memorial Gate, they paid tribute to the sacrifices made in wartime. Some 15,000 council homes and 20,000 private builds were added to Bristol's housing stock in the interwar period. Demobilisation of the armed services and the decommissioning of much of the defence infrastructure, was accompanied by the less welcome lay-off of many female employees. All this ushered in a brief post-war boom which was soon to be followed by a deep economic recession. Defence spending plummeted and the government adopted the principle of the Ten-Year Rule which severely restricted defence spending on the basis that it would take ten years for a potential enemy to rearm sufficiently to present a threat to national security. Not until the 1930s would efforts commence to compensate for the neglect of the previous decade. Ironically in the midst of this official retrenchment, civilian flying boomed, stimulated by the exploits of, amongst many others, Amelia Earhart, Amy Johnson, and Charles Lindbergh. Flying clubs started up and the Automobile Association laid out a countrywide network of landing grounds for the use of their members. Many of these were simply flat fields with a tank of fuel in one corner, while others ran to a hangar as at the old Great War airfield at Leighterton. Others in the county were outside Tewkesbury, Cirencester, and about a mile north of what would later become Staverton airfield.

The Armed Forces on the Eve of Another War

The Regular and Territorial Armies

In 1920, the Glosters' 1st Battalion was deployed to County Cork for internal security operations seeking IRA members and hidden arms caches and in 1927, the Glosters' 2nd Battalion went to China as the British element of an Indian Army brigade sent to safeguard the International Settlement in Shanghai against Japanese expansion. Come 1939, and the 1st Battalion was in Rangoon, Burma, and the 2nd Battalion was in the 3rd Infantry Division at Stonehouse Barracks, Plymouth. The National Shipyards had never got off the ground, and in 1924, that at Beachley Point was put to a different use. The Central (Army) Training School for boys had been set up at Aldershot the previous year, but now moved to Beachley, to be accommodated in the former POW Camp, and was renamed the Army Technical School (Boys) in 1929. Families of the school's staff lived in the houses put up for shipyard workers in Sedbury. The TF, suspended for the duration of the war, was reformed in 1920 as the Territorial Army (TA). In 1926, a TA Camp was held at Stroud, but swingeing budget cuts meant that such activities would be curtailed until well into the 1930s. Although the TA was given air defence as its primary role, little was done to either train or equip it for that onerous task which demanded a high degree of technical skill and the provision of complex equipment. The RGH Yeomanry had exchanged its horses for armoured cars by 1923, initially the Peerless, armed with two Hotchkiss machine guns and then, after 1930, old Rolls-Royce models which had been developed for the RNAS in 1914, and seen service with Lawrence of Arabia. Each squadron maintained a couple of them and then drew the balance from a central pool for summer camps. As the need for armoured units was gradually becoming better appreciated at the War Office, infantry as well as yeomanry units were being converted to armour, thus the Glosters' 6th Battalion (TA) became 44th Battalion, Royal Tank Regiment. Similarly, the TA's role as the principal AA force saw other such conversions when the Glosters' 4th Battalion became the 66th (Glosters) Searchlight (S/L) Battalion, RE, then Regiment, RA. Fitful rearmament had begun in the early 1930s, but it was not until 1938 that much was done to upgrade the army. There was little scope to do much with the regular units, so it was the TA, with room for expansion, which saw most change. Each TA unit was required to clone itself by using cadres of officers, NCOs, and other ranks as the basis for a parallel new unit. The RGH produced a second-line regiment, with its HQ on College Lawn, Cheltenham, and squadrons in Stroud and Tetbury, while the 1st RGH remained in Bristol, Gloucester, and Cirencester. The 2nd RGH was still recruiting at Tetbury's Ormand Head Hotel into November 1939. The 1st RGH was equipped with Vickers light tanks, but the only tracked vehicle that could be found for the 2nd RGH at this time, was a solitary Dragon artillery tractor. Left as a lone TA battalion, the Glosters' 5th Battalion opened a new drill hall in Gloucester's Painswick Road from whence it cloned a 7th Battalion in August 1939, with its HQ in Stroud. Recruits came from across the northern half of the county with some signing on at Cirencester's Bingham Hall during the July recruitment drive. Also recruiting in Cirencester in 1938 was 23 LAA Regiment, a completely new unit while, straddling the outbreak of war, new TA Centres opened at Horfield and Bedminster.

Cheltenham, 2 College Lawn, HQ of the 2nd Royal Gloucester Hussars Yeomanry.

Gloucester, the Painswick Road drill hall of the 5th Battalion, Glosters.

On the eve of war, all the Gloucestershire units came under Southern Command apart from the Beachley training school which was administered by the Welsh Area of Western Command. The 1st RGH was part of the 20th Light Armoured Brigade with its HQ in Gloucester. The 44th RTR remained based in Bristol but, operationally, was part of the 21st Army Tank Brigade based in Chipperfield near Watford, while its duplicate 50th RTR also remained based in Bristol but belonged to the 23rd Army Tank Brigade in Liverpool. Horfield Barracks remained the Glosters' regimental depot; their 5th Battalion was in one of the 48th Infantry Division's three brigades; and the new 7th Battalion similarly joined the second line 61st Division. There were also two RA Survey Regiments (TA), a RE Field Company in the Forest of Dean area, and a detachment of a RE survey battalion based in Bristol. The 46th (Gloucestershire) AA Brigade, also based in Bristol, consisted of the regular 76 and 98 HAA Regiments and 66 S/L Regiment (TA) with RASC and RAOC support. The important Aircraft Storage Unit (ASU) at RAF Aston Down, was defended by 131 LAA Battery RA.

The militia was reintroduced in 1939 when all twenty- to twenty-two-year-olds were conscripted for six months' full-time military training, based in existing barracks and in hutted camps specially constructed for the purpose. In Gloucestershire, there appear to have been four locations for this militia training: Horfield Barracks; Reservoir Camp in Gloucester which was built in 1939; Cirencester Park, later to become an American hospital; and Spring Lodge militia camp, near Chipping Campden, later to be a US Army and POW Camp officially known as Springhill Lodges accessed by a private railway siding in Moreton-in-Marsh. Men registered at their local Labour Exchange, but by the time that the first intake had barely enrolled, war had been declared and full-scale conscription of a much wider age group proceeded. Eight companies of the Auxiliary Territorial Service (ATS) were in process of being raised in Gloucestershire in January 1939. Their members would be attached to local TA units and work as drivers, cooks, storekeepers and clerks. Volunteers needed to be able to get to Cheltenham, Gloucester or Bristol for weekly drills. Other women, not wishing to join the ATS might opt for the Women's Land Army, originally formed in the previous war, with training courses being run by the Royal Agricultural College at Cirencester.

In 1923, the Bristol RNVR got a new drill ship—HMS *Flying Fox*, a minesweeper launched in 1918. It had been refitted as a HQ ship at Pembroke Dock and was moored at Mardyke Wharf. Its silhouette was transformed by the removal of most of its superstructure to accommodate a drill hall, but the guns and lifeboats were retained. In 1939, seven RNVR Divisions were being recruited in the county.

The RAF Prepares for War

The aftermath of the Great War saw the majority of the RAF's aircraft scrapped and most of its airfields closed and returned to agriculture. Despite having to fight for its very survival, the newly minted RAF was able to make a case for gradual expansion. As Germany replaced France as the anticipated opponent, and the strategic planners looked to expect aerial attacks from the east, then the RAF allocated its various operational and support functions across the country. Gloucestershire was considered a safe place for training pilots at Filton, Staverton, Little Rissington, and South Cerney; for storing aircraft at Aston Down and Little Rissington; and supplies and equipment at Quedgeley.

Filton had begun to train pilots before the Great War and after the war the Bristol company was contracted by the Air Ministry to run a Reserve training school for the RAF opening in 1923 and initially flying a range of Bristol aircraft, but standardising onto DH Tiger Moths in 1933. A Reserve squadron had also been formed in 1929. In 1937, the school was integrated into the nationwide training regime but continued under Bristol's management becoming No. 2 Elementary & Reserve Flying Training School (E&RFTS). The Reserve squadron joined the RAF in 1938 as 501 Squadron flying Hawker Hurricanes and No. 2 Ferry Pilots' Pool was formed to fly out the Blenheim bombers being built in the Bristol works. Little Rissington was built in 1937 by Wimpey's whose contract was worth well over half a million pounds (£25 million in current terms). It was a purpose-built training airfield, opening in August 1938, for No. 6 Service Flying Training School (SFTS) which arrived from Netheravon (Wiltshire), and sharing space with No. 8 Maintenance Unit (MU) which stored aircraft for distribution to operational units. By 1935 it had become obvious that Germany was the likely enemy in any forthcoming conflict and that the eastern side of England would be no place for novice pilots. No. 3 SFTS had operated from RAF Spittlegate, outside Grantham (Lincolnshire), since 1928 but it was decided to find it a safer haven. Building commenced at South Cerney in 1936 for a new, purpose-built training airfield which opened in August 1937, enabling the training of Short Service Commissioned Officers to begin in 1938, initially flying the Hawker Audax and Hart and then, as they became available, the Airspeed Oxford. Staverton had opened as Gloucester's municipal aerodrome in 1936, and the RAF agreed to extend the airfield and to install night-flying equipment, sealing a twelve-year contract to use the airfield for training purposes. In September 1938, Surrey Flying Services commenced pilot training as No. 31 E&RFTS and were followed in May 1939 by Airwork Civil School of Air Navigation who also trained RAF observers using Dragon Rapides. Rotol used many aircraft for airscrew trials, beginning in April 1939 with a Wellesley bomber. The aerodrome became RAF Staverton officially in September 1939 and No. 31 E&RFTS closed. The Gloucestershire skies were busy night and day during this time of preparation. In April 1939, gliders were being flown out of the old AFC airfield at Leighterton and night-flying training was taking place at South Cerney. The RAFVR HQ at Carrick House, Lypiatt Row, Cheltenham, began intensive recruiting for all relevant trades in April 1939 with widespread advertising in the local press. Aston Down, utilising the site of the former Minchinhampton airfield, opened in October 1938 as No. 7 Aircraft Storage Unit (ASU), and was immediately renamed as No. 20 MU and used for the storage of equipment. Nailsworth's George Hotel provided officers' quarters until sufficient accommodation had been built. When aircraft began to arrive, it was so they could be prepared for shipment abroad. RAF No. 40 Group (Equipment) formed in January 1939 and opened RAF Quedgeley, No. 7 MU, as one of seven national equipment depots, on six sites centred on the former Great War munitions factory.

The three Gloucestershire airfields constructed during the RAF's expansion period of the 1930s: South Cerney, Aston Down, and Little Rissington, all adhered to a set of nationally agreed standards, which enjoyed the approval of the Council for the Preservation of Rural England (CPRE) and the Royal Fine Arts Commission, steered by Edwin Lutyens. A neo-Baroque style of architecture was chosen to give these essentially rural developments the appearance and *gravitas* of grand country houses. All the buildings were designed by the Air Ministry's in-house architect, A. Bulloch FRIBA. Standard buildings performed regulation functions and were thus instantly recognisable. The four 'C'-type hangars at Little Rissington, and the three at South Cerney each form an arc on the edge of the grass flying field. Behind them stands an aircraft repair shed with accompanying main stores, main workshops, parachute store, armoury/photographic building, sickbay, MT section, airmen's dining rom, sergeants' mess, and two-storey barrack blocks. The T-shaped Station HQ at South Cerney stands near the main entrance opposite the Guardroom. Another very distinctive building is the officers' mess where two-storey residential wings flank a low central communal block but at Little Rissington this is an even more imposing three-storied building. Officers' garages and their indispensable squash court are adjacent to the mess. Terraces of airmen's married quarters, semi-detached NCOs' houses, and detached houses for officers, are grouped around the edge of the complex. It would take the first year of war to realise just how vulnerable was this nucleated design with everything concentrated in a single, densely populated area. Aston Down shares a similar suite of technical, administrative, and domestic buildings but its hangars, sensibly distributed around the site, are a mixture of types on six widely spread sites. The main site has a single 'C'-type for dismantling and erecting aircraft; two 'D'-types with bowstring roofs for storing complete aircraft; and five prefabricated Bellman hangars on steel frames, of two different heights: 25 feet (8 metres) and 17 feet (5 metres). Site C has two more 'D'-types. Site D has a pair of 'L'-type hangars, and Sites B, F, and G each have pairs of 'E'-type hangars, all 300 feet (92 metres) long and 150 feet (46 metres) wide with clear openings 40–70 feet wide. Both the 'E' and 'L'-types derived from the Lamella, a steel-framed German design, but the

'E'-type had both roof and supporting arches of reinforced concrete whereas the 'L'-type's reinforced concrete roof was supported on steel arches. Most of these hangars on sub-sites were given a covering of turf. In all, Aston Down had a total of eighteen hangars. Little Rissington, as well as being a training airfield, was also home to No. 8 MU and was provided with six 'E'-type hangars, holding almost 300 aircraft in July 1939, plus a complete squadron of Wellington bombers. The watch offices at Little Rissington and South Cerney are both recorded as being built to the same design but differ in several respects. The basic design (5740/36) for the chief flying instructor's block for use at Flying Training Schools comprises a central three-storey block with a single-storey projecting wing to each side. At Little Rissington, the central block is the standard three storeys in height, but the wings have been heightened to two storeys. At South Cerney, however, the wings are the usual single storey, but the central block is only two storeys high. The RAF supply depot complex centred on Quedgeley also shared the prevailing aesthetic, applied to its HQ and other administrative buildings such as guardhouses. The massive storage sheds with their distinctive clerestory roofs were built to a standard design from the same Air Ministry book of plans.

Aston Down airfield, a 'D'-type hangar on this Aircraft Storage Unit.

Little Rissington airfield, the Station HQ.

Little Rissington airfield, the watch office to a design of 1936.

Little Rissington airfield, a 'C'-type hangar.

Quedgeley RAF Storage Site, the HQ building.

Quedgeley RAF Storage Site, storage sheds.

Quedgeley RAF Storage Site, one of several guardrooms.

Air Defence

One of the more successful elements of the London Air Defence Area in the Great War had been the network of spotters, watching out for hostile aircraft and reporting their movements to a central control centre. This system was rolled out as Air Defence Great Britain (ADGB) in 1925, ultimately intended to cover the whole country. Although some Observer Corps posts in the south-eastern counties had been set up as early as 1924 in Kent, and in Essex in 1929, it would not be until 1937 that the network reached Gloucestershire: No. 23 Group, Bristol, with its HQ in Little King Street, at the rear of the public library; and No. 24 Group, Gloucester, with its HQ in the George Street GPO Telephone Exchange. Over the next two years, sixteen posts would be established from Avonmouth to Stow-on-the-Wold. These were grouped in threes and fours, linked by telephone landlines. Gloucester's Group L comprised North Cerney (24/L1), Andoversford (24/L2), Poulton (24/L3), and Cleeve Hill (24/L4). Observer posts, manned by volunteers on a rota basis, were basic. A timber shed offered some shelter and a rudimentary plotting instrument was introduced in 1935. These were not widely issued, particularly in what were considered 'back areas', a notion which would be made redundant by the German occupation of north-west France. Observers would report sightings of enemy aircraft to the Group HQ by telephone. Group would be in permanent contact with RAF Fighter Command which had established a sector station in Bristol, at Filton. On receipt of a warning, interceptor aircraft would then be scrambled. Recruits to the Observer Corps, preferably as groups, came from both municipal and commercial sources. Most training was predicated on the anticipation of hostile attacks taking place over eastern England and the operational area designated on the map of the 1938 Home Defence Exercises stops just short of South Cerney. A final exercise in August 1939 co-ordinated plots made by RDF (radar) and observers. Despite the Air Ministry's intention that observers would be uniformed, most would still be identified by only a badge and a brassard well into the war.

The Aircraft Industry

Formerly HH Martyn, the Gloucestershire Aeroplane Company bought what had been the Brockworth AAP with its flying field in 1928 and moved out of Cheltenham. It had quickly changed its name to the Gloster Aircraft Company (GAC) in 1926 as its new name was a problem for foreigners to pronounce. Taken over by Hawker Aircraft Ltd in 1934, it still continued to trade under its own name, its most significant product being the Gloster Gladiator, a biplane torpedo bomber used by the Fleet Air Arm, but by 1939, it was manufacturing large quantities of Hawker Hurricane fighters. The British & Colonial Aeroplane Company became the Bristol Aeroplane Company (BAC) in 1920 and acquired the Cosmos Engineering Company which built the Jupiter aero engine, designed by Roy Fedden, formerly of Brazil Straker, which had powered Bristol models in the past, and would continue to prove its popularity with many, mainly civil, aircraft manufacturers. Between

Filton, Pegasus House, the Bristol Aeroplane Company HQ.

1930 and 1937, the Bristol Bulldog was the RAF's primary fighter. By 1935, Bristol's Filton factory, covering over 60 acres (25 ha) and with a workforce of over 4,000, was building Blenheim bombers for the RAF. BAC's HQ building, the Art Deco Pegasus House, was opened by the duke of Gloucester in 1936. In 1931, George Dowty had set up Aircraft Components Ltd, in works in Lansdown Terrace Lane, Cheltenham. He then moved to Arle Court for a new factory with land for an airfield. The company made parts for the Gloster Gauntlet and then the Gladiator in 1934, and became Dowty Equipment Ltd. The key aeronautical development of the late 1930s was Frank Whittle's ground-breaking invention of the jet engine which would eventually revolutionise flight. Glosters was chosen as the company whose designer, George Carter, would design and build the airframe for this engine—the E.28/39. Other Gloucestershire firms benefitted from the rearmament programme as Wathen Gardiner Ltd began making uniforms for the RAF in 1937.

Fuel Storage

In 1936, there were only 8,000 tons of oil held in reserve storage and most of the nation's oil stock was held in the Thames Estuary or at ports around the country. To shift the balance away from the eastern side of the country, west of a line drawn from Southampton Water to the Tyne, and to increase exponentially the amount in reserve storage, new tanks were built to ensure that airfields might be assured an adequate supply. These stocks would be held at Aviation Fuel/Air Force Reserve Depots (AFRD), in C2 tanks, each holding 4,000 tons and capable of withstanding direct hits from incendiaries and near misses from HE bombs. Additionally, smaller, more numerous Air Force Distribution Depots (AFDD) would be built nearer to actual airfields. The AFDD would consist of four D1 tanks each holding 500 tons of aviation fuel and the first such tank was completed and filled in March 1939. Avonmouth was one of three key ports on the west coast for accepting imports of oil for distribution to these depots. It was envisaged that a reserve of 80,000 tons, a tenfold increase on existing holdings, would be sufficient but by 1940, that figure had been increased to 940,000 tons.

Preparing the Populace for War

Although the likelihood of conflict with Germany had begun to intensify from the mid-1930s, very little in the way of preparing the civilian population for a war where aerial bombing was anticipated from the start had been achieved. Emergency railway administration and control centres were initiated by GWR in Bristol and Gloucester in June 1938, but only the next year did real urgency kick in. Towns across the country began to recruit volunteers for Air Raid Precautions (ARP) and the Auxiliary Fire Service (AFS). In Fairford, a new fire brigade was set up. In February, the Women's Voluntary Service (WVS) for Civil Defence (CD) held its inaugural meeting in Cirencester, appointing village representatives who would prepare for such tasks as the placement of evacuees or the emergency feeding of refugees from bombed cities, and setting up an office in Cirencester's Dollar Street. A preoccupation with the threat of gas bombing put, as it happened, undue emphasis on the general provision of gas masks. In Cirencester, the Old Bathurst Museum was set aside as a store for the more than 23,000 respirators being assembled by forty volunteers. Each town identified the key locations in their ARP planning. In Cirencester, the ARP Control would be in the basement of the municipal offices and the first aid post at the Temperance Hall in Thomas Street. Thought was also given to the provision of air-raid shelters. Suitable buildings with robust basements were chosen but prefabricated models were also developed. Norcon Ltd's Concrete Pipe and Products Works at South Cerney was one of the companies involved, manufacturing prefabricated air-raid shelters from cylindrical concrete tubes, 4 feet (1.2 metres) long and 6 feet (1.8 metres) in diameter. Six of these could be joined together by watertight seals, forming communal shelters 24 feet (7.2 metres) long and buried under 3 feet (90 cm) of earth, seating thirty-five people. The same tubes planted vertically in the ground, and suitably loop-holed, would later serve as pillboxes one of which still stands outside Cirencester. In July, the ARP held a dress rehearsal in Cirencester, simulating bomb blasts and mobilising all their volunteers as wardens, firefighters, first aiders, and caterers. That same month, only weeks away from war, Gloucestershire was one of those areas selected by the Home Office to have its blackout arrangements tested visually from the air.

7

Gloucestershire and Bristol during the Second World War 1939–45

The county contributed to the nation's efforts on the Home Front and abroad in many ways. The Severn Ports were responsible for 10 per cent of British foreign and coastal imports, amounting to over 4 million tons in 1939. These were predominantly grain, petroleum products, dairy products, timber, and frozen meat. Despite the recession, Bristol's export trade had increased by 1939 to three quarters of a million tons. Avonmouth was also home to important industries such as Imperial Smelting's zinc works, while Filton and Brockworth represented vital elements of the aircraft industry. It would eventually be ascertained that the German 'Sealion' plan for the invasion of Britain required that everywhere to the south of a line drawn between Gloucester and Maldon (Essex) should have been secured by S+10. The German plan was to land 90,000 men in Lyme Bay, isolate Cornwall, and advance north towards Avonmouth and Bristol. Only in the summer of 1940 were the particular importance and vulnerability of the Severn ports formally recognised by General Headquarters (GHQ), who then ordered that the relevant defences be put in place by S+6. However, some measures to defend the county had been initiated from the very beginning of the war. The outbreak of war in 1939 had seen the mobilisation of the county's own units; the recruitment of extra units; and the arrival of a succession of units into the area for both short stays and longer ones.

The Deployment of Local Units

To avoid the problems encountered in the Great War when territorials, being legally only committed to service at home and therefore engaged on different terms to the regulars, the TA was subsumed into the Regular Army as soon as war was declared. While many new units had their roots in the existing establishment, the regulars also recruited from scratch.

Gloucestershire's Armoured Units

The 1st RGH, equipped with a mixture of armoured cars, light tanks, and Bren carriers formed part of a scratch yeomanry force based on Newmarket as a mobile anti-invasion force. By the end of 1940, armed with the unsatisfactory Valentine tanks, it served with the 20th Armoured Brigade in Kent until April 1943 by which time it had converted to an Army Reconnaissance Regiment with the whole brigade assuming a training role. From August 1944 until the end of the war it was the armoured element of the 55th (West Lancashire) Infantry Division and then, trained reinforcements for Montgomery's 21st Army Group in 35 Armoured Brigade. Its duplicate yeomanry regiment, the 2nd RGH, was part of 22nd Armoured Brigade and served in the Western Desert equipped with Crusader cruiser tanks, US Honey M3 light tanks, and the more heavily armed Grant tanks. The regiment suffered enormous losses in the retreat which finally halted the enemy advance at Alam Halfa and was disbanded in September 1942 but many of its personnel fought on, attached to other units at El Alamein and on through the campaign to clear the enemy out of Africa. From May 1944, the 3rd RGH was a skeleton formation using dummy tanks in deception operations in Egypt and Syria. Starting the war in Bristol, 44 Royal Tank Regiment (RTR), formerly the Glosters' 6th Battalion, was deployed to Egypt in early 1941 as part of 1st Army Tank Brigade and fought in all the major battles in the Western Desert. It then took part in the invasion of Sicily and fought its way up through Italy until January 1944 when it was recalled, to prepare for the Normandy Landings. For three weeks in March, this involved trialling some of General Hobart's 'Funnies', the specialised armoured assault vehicles of the 79th Armoured Division. Its duplicate 50 RTR spent the entire war in 23rd Armoured Brigade and fought through El Alamein, Libya, and Tunisia, up through Sicily and Italy, and thence to Greece via Palestine.

The Gloucestershire Regiment

In September 1939, following their six-week-long basic training at Horfield Barracks, newly called up men were housed in bell tents in Eastville Park, using water from the boating lake for all their needs. They were then sent to France to join the Glosters' 5th Battalion. The 1st Battalion was stationed in Burma on the outbreak of war and would be required to cover the retreat after the Japanese invasion of March 1942. Reduced by four-fifths of establishment, remnants of the battalion continued a fighting withdrawal, reaching India where it reformed but played no further active role. The 2nd Battalion landed in France with the BEF in October 1939 and was effectively wiped out in the retreat to Dunkirk losing over 400 killed and nearly 500 taken prisoner while buying time for those being taken off the beaches. Following complete reconstruction, the battalion returned to France on D-Day and had fought its way into Germany by VE Day. The 3rd Battalion was responsible for running the Regimental Depots at Horfield Barracks, Bristol, and Reservoir Camp, Gloucester, taking in recruits and carrying out their basic training before sending drafts to the other battalions. The 5th Battalion also fought with the BEF but managed to get just 400 men away on the boats from Dunkirk, returning to Eastville Park. Brought back to strength it carried out anti-invasion duties, but in 1941, it was converted into the 43rd (Wessex) Reconnaissance Regiment, while its duplicate, the 7th Battalion, served in Northern Ireland alongside the 9th Battalion. The 8th Battalion was formed from older soldiers who had hitherto constituted National Defence Companies and was employed in guarding VPs such as South Cerney airfield early in 1940. Following the reduction in the threat of invasion, along with the 9th Battalion, it had been broken up by 1942. The 10th Battalion was formed to serve as home defence troops and was assigned to 212th Independent Infantry Brigade. It served in South Wales in 1940–41, tasked with despatching a company-strength mobile column to Llandow airfield, where eight designated VPs are recorded. Later in 1941, it was on the Lincolnshire coast of the Wash at Holbeach. In 1942, however, it was converted to armour as the 159th Regiment, Royal Armoured Corps, serving as such in India for just six months. Its former identity had been resumed by the spring of 1943 and it fought in the battles in Burma as infantry up until VJ Day. The 11th Battalion had been created out of the regiment's Holding Battalion but was converted to 118 LAA Regiment in 1942. The 70th (Young Soldier) Battalion was formed as a temporary measure to train young men prior to service overseas, but though useful as home defence troops guarding VPs and airfields during the invasion scare, it was disbanded in 1942, having served its purpose. Together, the regiment's several battalions lost nearly 1,000 men killed during the war. The 66 (Glosters) Searchlight Regiment, formerly the Glosters' 4th Battalion, was given the task in 1939–40, of defending Avonmouth and Bristol. In 1942 the regiment relinquished its Market Street drill hall in Bristol to provide accommodation for American troops and moved into Badminton School in Westbury-on-Trym, where it remained until the summer of 1944 when it was deployed to London to defend against the V-1 offensive.

The County's Artillery Units

The elements which made up 46 AA Brigade (see below) all served in the UK for the first half of the war. Some were prepared for overseas service, one example being 76 HAA Regiment which was deployed to Tunisia in 1942 for Operation Torch while 98 HAA Regiment took part in the D-Day landings via Manchester. Initially attached to the 49 (West Riding) Infantry Division, 118 LAA Regiment then served in India where it was disbanded in August 1944, and 23 LAA Regiment went to Ceylon (Sri Lanka) in 1942.

One of those newly recruited units was 52 Heavy Regiment (RA) which had been disbanded after Dunkirk but was recreated in 1943 and is reported locally as being recruited in Thornbury. Initially armed with 9.2-inch howitzers of Great War vintage, by 1944 and fighting in northern Europe, the regiment was equipped with 155-mm Long Tom howitzers. In May 1941, the Maritime Anti-Aircraft Royal Artillery (MRA) was set up initially to defend ports against attack from the air. Its No. 3 Regiment, covering the south of England, had its HQ in Thornbury, along with No. 6 Battery which operated in the Severn Estuary. As aerial attacks on the ports diminished due to the German shift of resources to the Russian Front, these gunners were deployed to defend merchant ships at sea. The U-boat offensive introduced the need for heavier on-board guns to counter surfaced U-boats and 6 Regiment, MRA, based in the seventeenth-century Kyneton House, a mile to the north-west of Thornbury to serve ships in the Severn ports, set up a training school in the castle grounds and a range at Pilning on Severn Beach where LAA machine guns were fired out over the river. Gunners were trained and assigned to ships ranging from coasters to the great ocean liners which operated as troopships. Around 1,000 men went through the depot every month, and in December 1942, it had an establishment of over forty officers and 3,000 men. The MRA gunners operated in parallel with the DEMS base on HMS *Flying Fox* in Bristol, where training on the larger guns took place. Several buildings in Thornbury were taken over: the officers' mess was in the castle and the sergeants' mess in the picture house; the Exchange Hotel Assembly Rooms was the Quartermaster's Stores; a radio shop was the reception centre for men reporting between assignments; Alveston's riding school with huts on the Jubilee Field was a holding and training centre; further premises provided YMCA and other recreational facilities. Gunners served all over the world being particularly involved in serving on ships on Arctic, Atlantic, and Mediterranean convoys, with many going down with their ships.

The Armed Forces in Gloucestershire

From the outbreak of war, Gloucestershire would see a succession of units come and go. In September 1939, the 50th (Northumbrian) Infantry Division concentrated in the Cotswolds on its way to join the BEF in France in January 1940. While the Gloucestershire Hussars had become an armoured Yeomanry regiment, the Norfolk Yeomanry had become the 65th Anti-Tank Regiment, RA (TA) and was attached to the Geordies' division. The regiment found itself in Tewkesbury using a variety of stables, barns, warehouses, the Abbey Mill, the old drill hall, and a butcher's shop as accommodation. RHQ was in the Old Willow Café, and the officers' mess in the Hop Pole Hotel. They spent an uncomfortable winter alternately being flooded and frozen before being moved overseas in January 1940. The most frequent destination for drivers under training appears to have been Stow-in-the-Wold where the 4th Battalion of the East Riding Regiment spent the winter in tents, while the 5th Battalion, Green Howards was in billets in Chipping Campden. Just as the new Glosters' battalions were assigned to guarding VPs, so the 9th Battalion, Worcestershire Regiment, was defending RAF Little Rissington. While the Home Guard would later seek to enrol men of military age who were essential workers rather than, primarily, those who had retired, in July 1940, No. 1 Platoon, 'E' Company of the old sweats' National Defence Corps, stationed near Cirencester, was actively recruiting men from the forty-one to fifty-five age group at Bingham Hall, for guarding VPs in the locality. The 38th (Welsh) Infantry Division was a second-line TA division, formed in September 1939, deployed to the Severn Valley until May 1940, then returned for training to the Liverpool area. In July 1940, the 3rd Infantry Division, nicknamed the 'Iron Division' under Montgomery, was deployed in the Cheltenham area tasked with protecting Bristol from landings in South Wales or the Bristol Channel, and airborne landings in the Mendips. It was part of the GHQ Mobile Reserve and included the Glosters' 2nd Battalion. Also part of this Mobile Reserve force was 44 RTR within the 21st Army Tank Brigade. Having been evacuated from France in the May, the 42nd Infantry Division moved from East Anglia to Gloucestershire in November 1940 and units were dispersed across the area, the 5th Battalion, King's Own Royal Lancaster Regiment, part of 126 Brigade, being based in Winchcombe for some time. Like many of the TA divisions recovering from the traumas of their French experience, the division alternated between positioning itself for anti-invasion operations and training for the future. The 1st Battalions of both the Warwickshire Regiment and the Royal Ulster Rifles spent part of this period in Thornbury. The 55th Infantry Division which would spend the whole war in the UK, was deployed to the Thames Valley, replacing a division moving on. Its 164 Brigade, comprising three former TA battalions from Liverpool and south Lancashire, was deployed in the Lechlade–Cheltenham–Tewkesbury areas. Its HQ was at Bishop's Cleeve in 1941 with a Reinforcement Camp at Cheltenham Racecourse and an overflow at Swindon Hall. Daylesford House was home to the RASC and the Durham Light Infantry.

Some units were just passing through on their way overseas. The 2nd Battalion, Norfolk Regiment, was camped at Fairford by December 1941, prior to its embarkation to Singapore with the ill-fated 54th (East Anglian) Division. By early 1942, the 2nd Infantry Division, also recovering from the BEF operation, was stationed in the Bristol/Gloucester area prior to a move to India. Cirencester hosted an overseas unit when, in August 1940, the Stratton House Hotel became the HQ of the 11th Forestry Company, New Zealand REs, which operated sawmills in Hailey and Overley Woods, a few miles to the west of the town. Two years on and the county would see a 'friendly invasion' on a much larger scale.

The US Army in Gloucestershire and Bristol

In January 1942, the US Army's V Corps was the first formation to deploy to the European Theatre of Operations USA (ETOUSA), moving into mainly Southwest England having first landed in Belfast, establishing its Corps HQ at Clifton College. The four divisions comprising V Corps were employed in the North African landings known as Operation Torch but had returned to Britain via Liverpool in November 1943. Round about this time General Omar Bradley, who had created the 82nd Airborne Division and commanded several different corps as Eisenhower's deputy, was promoted and given command of the 1st US Army. This would spearhead the invasion of France under Montgomery's command. The HQ of V Corps had moved out to Taunton, so he established his own HQ in the now vacant Clifton College, living at a house called The Holmes, now located in the University of Bristol's Botanic Gardens, and commuting to Clifton where his staff was billeted. Planning for D-Day was centred on the school's library and committee room. The topmost storey of the Wilson Tower became a wireless room in direct communication with Eisenhower's SHAEF HQ at Bushy Park, Teddington; Montgomery's 21st Army Group HQ at St Paul's School in Hammersmith; the Chiefs of Staff Supreme Allied Command (COSSAC) in underground stations off Tottenham Court Road; and with GC&CS at Bletchley Park. Two of the five D-Day beaches had been allocated to 1st US Army which would land on Omaha and Utah.

The movement of US forces into Britain was designated Operation Bolero and the major concentration of this influx of US troops was in the south-west with a significant proportion coming to Gloucestershire and Bristol. Avonmouth provided convenient port facilities for the enormous build-up of vehicles, aircraft, gliders, ordnance, and equipment needed to supply the US forces bound for Normandy. The US troops in Bristol, Gloucester, and Cheltenham were mainly employed by the Service of Supply (SOS) organisation with most of the combat troops spread across the counties of Devon, Dorset, Wiltshire, and Somerset. Bristol was home to large numbers of QM, MP, and transportation units with port units based at Shirehampton. Engineer units were based in Highnam Court at Tewkesbury, Westonbirt, Stroud, Painswick, Tetbury, and Wickwar. Ordnance units lodged at Reservoir Camp, Gloucester, Woodchester Park, and Yate. An entire QM Truck Group was based in Gloucester and most of another at Pinkney Park, Tetbury. There were QM railhead units at Cirencester and Buckland; chemical units at Coleford, Wickwar and Bristol; postal units at Cheltenham and Stone; a bomb-disposal unit at Wotton-under-Edge; four whole Engineer Aviation Battalions were based at Great Barrington (Station 422); and Tetbury (Station 529) was the base for six QM Aviation Truck Companies. This vast logistics infrastructure was necessary for the supply, maintenance, administration, and support of the four divisions of the US 1st Army and the USAAF IXth Air Force and only a very few actual combat units were based in Gloucestershire. They included the 635th and 773rd Tank

Springhill Lodges, initially a militia camp and then a US Army base.

Destroyer Battalions at, respectively, Toddington and Maugersbury Park Camp, Stow-on-the-Wold; the 25th Armoured Engineer Battalion at the former militia camp at Springhill Lodge(s) near Chipping Campden; and the 44th Armoured Infantry Battalion at Bourton-on-the-Water. Chavenage House, near Tetbury, was used by US forces and tradition has it that Eisenhower was billeted there at one time, possibly during one of his tours of inspection in the south-west in the spring of 1944. The US 6th Armoured Division arrived in Britain in February 1944 and was installed at Daylesford House near Adelstrop and Batsford Park, Moreton-in-Marsh prior to its move to France in July 1944.

Training

The Glosters were based at Horfield Barracks with an overspill camp in Eastville Park but in 1940, their regimental depot moved to the reservoir camp in Gloucester, the former militia camp which had opened in 1939. From the beginning of the war, each individual regiment had trained its own recruits at its own regimental depot, but this system was regarded as wasteful of instructors, so in the second half of 1941, a reduced network of twenty-five Infantry Training Centres (ITC) was set up. In August 1941, No. 15 ITC was established at the Reservoir Camp, providing basic training for new recruits to the Glosters, the Somerset Light Infantry and the Wiltshire Regiment, moving to Colchester in 1942. After its disastrous experience in France, the Glosters' 2nd Battalion was rebuilt and carried out a range of home defence duties based at Reservoir Camp. The training schedule included guarding the Gloster aircraft factory at Brockworth and Badminton House where Queen Mary had settled for the duration. It was brought up to operational readiness in time to take part in the D-Day landings, but a depot party of the Glosters remained at Reservoir Camp throughout the war. The published byelaws for the New Passage ranges at Redwick, on the left-bank of the Severn and established in 1935, and now the Pilning Rifle Range, describe them as for the use of A/T rifles, machine guns, mortars, revolvers, and grenades.

Prior to joining the BEF in France, the 50th Division carried out an exercise in January 1940, involving the whole division in an advance up the Northleach to Cirencester road, reacting to air attacks by the RAF and establishing a defensive position. This was a rare occurrence when so much training was piecemeal involving only small units. Much of the army would spend the four years between Dunkirk and D-Day on anti-invasion duties and in training. The 22nd Dragoons, raised as one of six entirely new armoured regiments in 1940, was part of 30th Armoured Brigade, from January 1941 until August 1945, initially in the 11th, then the 42nd, and, from October 1943, in the 79th Armoured Division. This was General Hobart's innovative formation developing a range of specialist AFVs dubbed 'The Funnies'. The 22nd Dragoons were one of the division's regiments using adapted Sherman flail tanks, known as Crabs, to clear paths through minefields by exploding the mines with the revolving flail, 8 feet (2.5 metres) wide, attached to the front of the tank. They were based at Maughersbury Manor, Stow-on-the-Wold, and trained on the cricket field, learning to operate trios of tanks in staggered line-ahead, or echelon, formations to the horror of local cricketers. Training, especially in the build-up to D-Day could take place in a variety of settings. The army trained on aircraft factory land at Yate, and Shenbourne Park, Cheltenham, accommodated the 51st (Highland) Division's Battle School from 1942, providing a number of activities including scaling cliffs. US troops used the lakes at Woodchester Mansion to train for the D-Day landings.

The Royal Navy in Gloucestershire and Bristol

Highnam Court was requisitioned by the Admiralty as a new-entry recruit training centre designated as a tender to HMS *Ganges*, and from 1942 to HMS *Cabbala*, at Tortworth Court, Falfield, with accommodation for up to 3,000 trainees taking the wireless telegraphy course. While officers and POs were billeted in the mansion, most trainees were in tents and when the grounds were flooded in 1942, the school moved out. Tortworth Court, now a hotel, had been requisitioned by the Admiralty in 1939 as an officer training establishment but specialised in coding and signalling and moved to Lancashire in September 1942. Huts had been added in the grounds and these formed the basis of the US hospital which then moved in, and of HMP Leyhill after the war. Like all the services, the Admiralty was desperate for storage, requisitioning the iron framed Standley Cloth Mill of 1813 at King's Stanley. The ammunition depot in Hawthorn Tunnel, Daybrook in the Forest of Dean, and the Britannia and Victoria Warehouses at Gloucester Docks were provided with railway sidings which linked to the local lines. The Bristol RNVR depot ship, HMS *Flying Fox*, served as a training ship for naval gunners on Defensively Equipped Merchant Ships (DEMS) until 1945, working in concert with their army colleagues of the MRA at Thornbury. A sea training ship, the TS *Vindicatrix*, had been moored up in the Thames to serve the Gravesend Sea School and was evacuated to the Sharpness Canal in 1939

Cheltenham, Shenbourne Park, 51st Highland Division Battle School, 1942.

with a hutted camp, traces of which can still be found, at Sharpness Dock. Youth organisations were strongly supported by the county's youngsters, and Cirencester's Sea Cadets paraded in Cheltenham as part of a 320-strong local detachment and also attended a camp at HMS *Impregnable*, Devonport, in 1940.

Anti-Invasion Preparations

Following the disastrous campaign in France and the Dunkirk evacuation, the county readied itself for invasion in the summer of 1940. As the Sealion blueprint would later confirm, in the event of Army Group C's plans being implemented, a successful landing in Lyme Bay would be followed by a speedy advance north through Dorset and Somerset and, within a week of landing on British soil, the capture of Bristol, Avonmouth, and Gloucester. British response to these perceived threats, in the absence of any mobile capability following the abandonment of so many tanks, guns and vehicles in France, would have to be an interim solution, quickly embraced by General Ironside. This was the Coastal Crust comprising coast artillery batteries, anti-landing obstacles, minefields, pillboxes, and infantry fieldworks. Inland from this would be a linear barrier known as the GHQ Line, stretching from the Bristol Channel at Highbridge right across the country into Kent, then up through the eastern counties to Yorkshire and beyond into Scotland. It would consist of continuous lines of A/T obstacles using rivers, canals, railway embankments or cuttings, and, where necessary, A/T ditches, all reinforced by pillboxes, A/T gun emplacements, minefields, and weapons pits. The purpose of these two major defence lines, supported by a network of similar but shorter lines, was to delay the enemy to gain time for any mobile troops being held in reserve at strategic points to converge on the invaders. Additionally, the Taunton Stop Line ran from north Somerset down to the east Devon coast, effectively cutting off Devon and Cornwall. Along with other lines running from Bath through Salisbury to Bournemouth, these represented a strategy of channelling an invading force into narrow corridors, thus rendering it vulnerable to constant flank attacks.

The Coastal Crust

In the absence of reliable information as to where either raiding in force or full-scale invasion might strike, the Coastal Crust extended around much of Britain. Given the importance of the Severn ports, both banks of the Severn were fortified from Beachley Point up to Arlingham. Survivors of the many more originally built include two Type 26 pillboxes at Beachley Point, constructed in the local fashion using pre-cast concrete panels as formwork; an octagonal pillbox, at Lydney Harbour, standing on the flood bank, again with pre-cast panels, with low-level entry and eight loopholes; and three rectangular pillboxes of similar

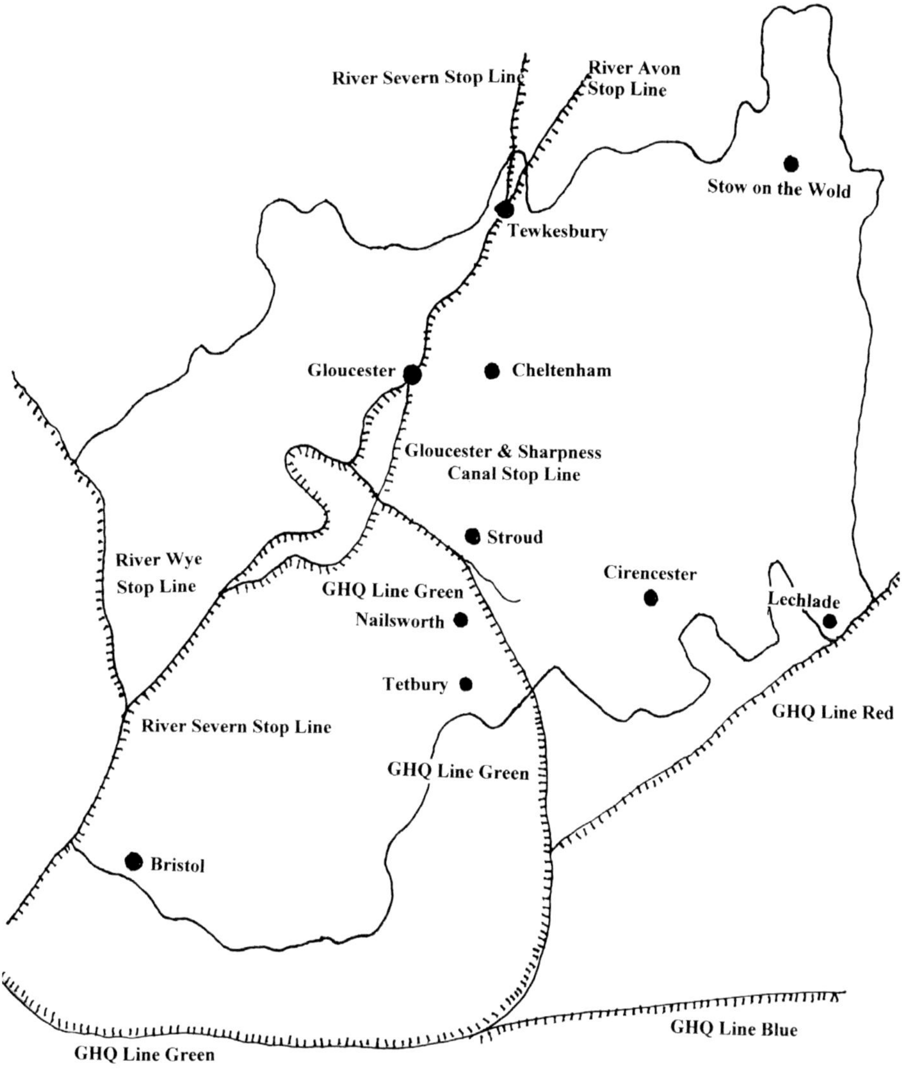

Sketch map of Gloucestershire to show stop lines and nodal points.

Beachley (ST547903), a pillbox to drawing number DFW3/26 with prefabricated concrete panels bolted together as formwork for poured concrete.

Arlingham (SO701121), a pillbox on the left bank of the River Severn.

construction, between the riverbank and the railway at Woolaston. Recent (2021) building operations at the inn at Beachley have revealed a pillbox overlooking the ferry slipway. On the east bank of the Severn, more rectangular pillboxes line the riverbank, four at Slimbridge with others at Tites Point, Purton, and at Arlingham. One pillbox at Arlingham was built into the prow of a beached barge, jutting out into the river, but appears now to have been washed away. These east bank pillboxes were built of vertical concrete slabs, tightly butted together with mortared joints. Their roofs were supported on concrete beams and covered with earth to promote a turf disguise.

The GHQ Line: Green and Red

The western end of the GHQ Line which curved around Bristol was the GHQ Green Line, or Outer Bristol Defences creating, in effect, a large Anti-tank Island, with all approaches to the city and its hinterland blocked. This line followed the River Brue from Highbridge to Wells; then the Somerset & Dorset railway and the Wellow Brook to Bradford-on-Avon; and thence the River Avon to Malmesbury. At Long Newnton, the line entered Gloucestershire, continuing past Tetbury and Avening, descending to Nailsworth and then via the Stroudwater Canal and the River Frome, terminating at the Severn at Upper Framilode. The main GHQ Line continued its progress eastwards by branching off from the Green Line to form two parallel defence lines which would rejoin at Reading. From a few miles east of Bradford-on-Avon, the GHQ Blue Line ran along the Kennet & Avon Canal, and from a point south of Malmesbury, the GHQ Red Line followed the Thames.

The GHQ Green Line in Gloucestershire relied on a range of defensive elements. From Long Newnton, astride the Fosse Way, to Avening, is a stretch of line where problems caused by the geology had been anticipated in the RE Reconnaissance Report and reiterated in a document issued by the commander of 48th Infantry Division in the first week of July 1940 which confirmed that the REs were arranging the necessary work with civilian contractors, McAlpine & Co., and County Surveyors. In fact, the A/T ditch was successfully dug, a total length of 7.5 miles (12 km) between Malmesbury and Avening and is still traceable. Crossing points were either removable plank bridges or, if permanent such as a railway bridge, then provided with concrete A/T cylinders to be rolled into place if necessary. All these ditches and obstacles were defended by shell-proof hexagonal Type 24 pillboxes. Between Stroud and Nailsworth, a steep-sided gorge supplied an effective natural barrier to tanks. The canal and river between Stroud and the Severn, were reinforced by more shell-proof Type 24 pillboxes constructed, unusually, as were those below Avening, using pre-cast concrete slabs, bolted into place, and filled to a thickness of 42 inches (1.07 metres), a design not seen outside the county. Another local feature was the L-shaped covered porch, apparently a later addition in brick, protecting the entrance. Apart from the mainly Type 24 shell-proof pillboxes, there was a pair of square splinter-proof Type 26 pillboxes at Long Newnton, one of which lies in Gloucestershire. At Bond's Mill, Stonehouse, alongside the canal, is a very unusual octagonal double-decker pillbox, now known as the 'Gatehouse', built to defend the Sperry gyroscope works. The lower level is of reinforced concrete while the upper is of brick with a flat concrete roof on which, reached by a fixed steel stair to a roof-hatch, is a position for a LAA gun mounting. On one side is a projecting porch supporting an external stairway to the upper level. Windows, inserted at both levels post-war, have largely replaced loopholes but three, with concrete lintels, remain at the higher level. All this work was carried out between May 1940 and early 1941 when the policy changed, and the construction of hardened defences, other than at airfields, ceased.

The GHQ Red Line formed a rear support for the main Blue Line. In Gloucestershire, it ran along a short length of the Thames to the east of Lechlade towards Kelmscott. Here, shell-proof Type 22 pillboxes lined the riverbank at regular intervals. Five of them straddled the weir at Buscot, with a sixth at St John's Bridge. These pillboxes were constructed to a design put out to contractors via supervisory RE officers and are peculiar to those defences put up by V Corps. In an attempt to save on materials, the front five walls of a standard Type 22 regular hexagonal pillbox were furnished with Bren gun embrasures and thickened to 42 inches (1.1 metres) while the rear wall remained at 15 inches (40 cm). To minimise the risk of flooding as they sit on the edge of the Thames, they are all raised on massive plinths. They were constructed by pouring reinforced concrete into timber formwork with prefabricated concrete loopholes added in as the build reached the appropriate height. The roof was poured onto corrugated iron sheets supported on the Y-shaped interior anti-ricochet wall. This would suggest an early date around midsummer 1940, since although proximity to Gloucester's timber wharves might have helped, timber generally became more difficult to source, and many subsequent builds used bricks for shuttering. In addition to the static defences, the Home Guard mounted a river patrol of armed launches the length of the Thames from Lechlade to Teddington Lock in west London.

A further defence line, partly in Gloucestershire, ran along the River Wye from Chepstow to Hay-on-Wye. Bigsweir Bridge was heavily defended with rectangular pillboxes and an L-shaped section post, with another similar structure covering the railway bridge at Sedbury on the Beachley Peninsula.

Right: Stonehouse (SO794053), Bonds Mill, an octagonal two storey pillbox with a LAA emplacement on its roof.

Below: Tetbury (ST889971), Star Farm, a hexagonal, shellproof pillbox to drawing number DFW3/24 with the local characteristic of an added porch, on the GHQ Green Line.

Long Newnton (ST918915), a square, splinter-proof pillbox to drawing number DFW3/26 on the GHQ Green Line, also with a porch.

Buscot (SU235982), River Thames, a hexagonal shellproof pillbox raised on a plinth but with a thinner rear wall to save on materials (and hoping the enemy will not come round the back).

Anti-tank Islands, Nodal Points, and Centres of Resistance

Initially complementing these linear systems but later offering an alternative, networks of strongpoints known as nodal points were created. These were generally locations where road or rail routes met, often at river-crossings. Tetbury, Stroud, and Nailsworth were designated as nodal points with all-round defences based on defended localities, often centred around roadblocks, and with central, last-ditch, defences termed keeps. These defences relied heavily on barbed wire, weapons pits, A/T obstacles, and mines. For a year or so after its return from France, 144 Infantry Brigade, which included the 5th Glosters, oversaw the construction of the GHQ Green Line under the direction of 224 Field Company RE, and was subsequently responsible for its manning. Later in 1941, the army stopped manning the GHQ Line altogether and shifted its focus to nodal points, but the Home Guard was permitted to man those stretches of GHQ Line which traversed their home villages.

Bristol, inside its Green Line defences, was classed as an Anti-tank Island, a self-sufficient garrison, whose HQ was in St Mary's Tithe Barn in Shirehampton High Street. Troops at Horfield Barracks, the AA units defending the city, and the various logistics units all came under the orders of the garrison commander. He worked very closely with the Home Guard Zone commander who controlled nine battalions as well as anti-aircraft and transport units. Joint exercises were held to test the city's defences and the efficiency of the garrison's cooperation with the city's Civil Defence organisation, particularly with regard to decision making and communications. In addition to the three strongpoints of Tetbury, Nailsworth, and Stroud relating to the GHQ Green Line, there were more at Gloucester, Tewkesbury, and Cirencester. A draft paper prepared by Southern Command in August 1941 listed Centres of Resistance which were seen as places that could impede enemy progress using obstacles and delaying tactics. Tetbury had been reduced to this lesser category and was garrisoned by a company of 125 men of the 6th Battalion, King's Own Yorkshire Light Infantry. Other locations in the South Midlands Area so defined were Lechlade, Stroud, Tewkesbury, Stow-on-the-Wold, and Cheltenham. Further papers issued in November 1941 confirmed those places still defined as Anti-tank Islands. Bristol remained in the higher priority category with Gloucester and Cirencester included in the lower. It was expected that the local authorities would ensure that these three A/T Islands were provided with adequate food stocks, firefighting and road clearance personnel, and facilities for dealing with battle casualties. Within three months, A/T Islands had become nodal points and a three-tier hierarchy had evolved. Bristol remained in Category 'A', but Gloucester and Cirencester had been demoted to Category 'C'.

Bureaucratic semantics aside, all these places had roles to play in the defence of the nation. Tewkesbury is where the Severn and the Avon part company, each to provide the basis for the defence lines which would ring the industrial heartland of Birmingham and the West Midlands. The Tewkesbury A/T Island included a roadblock on the bridge over the Avon; a large circular pillbox and associated A/T

Shirehampton High Street, the barn which constituted HQ Bristol Garrison.

Above left: Tetbury (ST888932), west side of New Church Street, five slots for A/T Rails or 'hairpins', sunk into the pavement.

Above right: Mickleton (SP162436), a loopholed wall commanding a bend in the road. (*Adrian Armishaw*) obstructions at the road junction south of the railway bridge; another pillbox, later demolished by US troops accessing the Ashchurch vehicle depot; and a hexagonal pillbox opposite the exit of the A438 road from the town bridge. Cirencester's defences, too, have been much reduced, but pairs of loopholes can be seen in the wall of Cirencester Park and in a wall on the Burford road out of town. A Norcon pillbox, little more than a sewer pipe with loopholes, has been repositioned at the west end of town. The railway, now disused, was protected by A/T blocks at Watermoor and a mobile spigot mortar mounted on a gangers' trolley. Local Home Guard commanders could fashion their own defences around the tactical plans for defending their village. At Mickleton, there are four loopholes cut into an old stone wall near the church, covering a bend in the road. Winchcombe was not listed as a Nodal Point but is nevertheless surrounded by a ring of pillboxes disguised with outer skins of Cotswold stone. Other defences included flame fougasses designed to shower the enemy with flaming liquid projected from buried roadside oil drums.

The Defence of Vulnerable Points (VP)

Local army commanders at district, area, and command levels retained detailed lists of VPs. These included transport and communications infrastructure, public utilities, key industrial concerns, and military and civil targets. Initially these were guarded by the same TA units which were charged with safeguarding airfields, but the task soon fell to the Home Guard. Fears of a land-based invasion were accompanied by the threat of parachute drops or troops being landed by transport aircraft. Certain particularly vital targets were the subject for concern at the highest level. The Gloster Aircraft factory was the focus of inconclusive correspondence between Dowding of RAF Fighter Command and General Kirke of GHQ who enquired as to the procedures for the (R)OC informing the relevant Army Command of parachute drops. Airfields, as we shall see, were singled out for particularly secure defences, but any flat space, capable of offering a clear landing area had to be identified and obstructed. Tewkesbury's water meadows were judged capable of use by transport aircraft or gliders and suitably obstructed by grids of ditches and earthwork obstacles. Fields had old cars and farm carts filled with earth or rubble scattered across them; arterial roads were strung with cables between high trees; and even flat beaches were furnished with forests of poles. Flat stretches of sheltered water were seen as possible landing places for seaplanes and were seeded with obstacles or covered by machine-gun positions. Airfields under construction had their runways obstructed. Odd pillboxes crop up on long-abandoned searchlight sites which had once been envisaged as strongpoints, or on road junctions. Pillboxes around Winchcombe were camouflaged with a stonework skin. One square brick-shuttered pillbox by a railway bridge on Gypsy Lane near Filton was offered free to a good home recently as it was proving an impediment to progress. Bristol's War Room in the basement of 19–21 Woodland Road, Tyndalls Park, is guarded by a pillbox on the street corner. Two square Type 26 pillboxes

Bristol, Tyndalls Park, the pillbox defending Bristol's War Room in the basement of 19–21 Woodland Road, built into the garden wall.

Saul (SO746085), a DFW3/26 pillbox on the towpath where the Gloucester & Sharpness Canal crosses the GHQ Green Line.

Greet (SP023303) near Winchcombe, one of several local pillboxes camouflaged by cladding them in stone.

remain on the Sharpness Canal. They are both constructed from the same pre-cast concrete slabs bolted together. Each has four loopholes and a brick porch with a concrete roof. One is at a road bridge at Saul and the other is where the canal crosses the GHQ Green Line at Saul Junction.

Airfield Defences

Recognising the threat to airfields from either paratroops or air-landing forces, the Air Ministry invited General Taylor, director of Fortifications and Works at the War Office, to carry out an assessment of their defence needs and to formulate a policy for implementing his recommendations. Taylor determined that airfields within 20 miles (32 km) of a port defined as 'vulnerable' should be placed in the top priority and receive the highest level of defence provision. Filton's proximity to Portishead and Avonmouth thus qualified it as a Class I airfield entitled to twenty to thirty pillboxes, both inward and outward facing; three Pickett-Hamilton 'disappearing' forts; a garrison of 274 troops to man the pillboxes; eight AA gun-pits; four extemporised armoured vehicles and numerous slit trenches, all suggested in Taylor's report. The next highest category of priority contained those airfields deemed necessary for counterattacking an invader, such as Moreton-in-Marsh, still under construction, but included in Class IIa. Aston Down and Little Rissington, both hosting the precious reserves of aircraft in ASUs, came into the IIc category. These Class II airfields required fifteen to twenty-four pillboxes, fewer AFVs, and a garrison of 225 troops. The residue of the nation's airfields, including South Cerney, fell into Class III where ten to sixteen pillboxes were thought sufficient with a garrison of 191 troops. Several of the county's airfields were either factory fields as was Brockworth or were no more than a gleam in the Air Ministry planners' eyes. The garrisons were mainly supplied by third-line TA battalions such as the 8th Glosters at South Cerney and Pucklechurch, or the 9th Worcesters at Little Rissington. Younger members of the Glosters' 8th Battalion were formed into the 70th Young Soldier Battalion and guarding airfields became a vital part of their training regime. Each airfield was allocated fifty rifles for arming RAF ground personnel, often, under the command, for no discernible reason, of the station's meteorological officer.

Pickett-Hamilton Forts were ingenious but horribly dangerous. The fort consisted of two concentric concrete cylinders, sunk flush with the ground surface. The inner tube could be raised by either hydraulic or pneumatic means to reveal machine gun embrasures about 18 inches (45 cm) above ground level. Three of the forts would be positioned at runway intersections, remaining in the lowered position until enemy transport aircraft began to land when they would be raised into the firing position. When lowered, all that was visible was a circular concrete apron with a small, square entrance hatch. As a Class I defended airfield, Filton was included in the sixth tranche, due to have three forts installed in February 1941.

The majority of the pillboxes constructed for airfield defence were the hexagonal, splinter-proof Type 22, two of which remain at Aston Down, one in a fragmentary state. In 1941, by which time airfields were the only recipients of pillboxes, the Air Ministry decreed that they should be thickened to 54 inches (1.4 metres) and the number of loopholes should be halved. It is doubtful if many of those airfields listed by Taylor ever received their full complement, although records show Little Rissington as having a generous fourteen concrete pillboxes and a further fourteen brick blockhouses, of which at least one survives. But only a very few examples of Gloucestershire's airfield defences survive, and these appear to reveal an eclectic and haphazard range of provision. Stoke Orchard has a boomerang-shaped Seagull Trench, while Windrush has an octagonal pillbox with an open, central well for an LAA machine gun, surrounded by a lateral passage accessing loopholes in seven faces, and Staverton retains several pentagonal pillboxes with four loopholes and an entrance in the fifth face protected by an open well with a mounting for an LAA machine gun. Southrop had an unusual structure at each end of the airfield where circular pillboxes are added onto the forward angles of a rectangular blockhouse. Adjacent to each of these structures stood a Norcon-type roofed pillbox. The Air Ministry designed a standard Battle HQ as an aid to co-ordinating an airfield's defence. This was in the shape of a suite of underground chambers for PBX, messengers, etc. with a concrete cupola raised up on low pillars above, affording the defence commander an all-round view of the landing field through the slit. Examples remain at Stoke Orchard, Chedworth, and, greatly overgrown, at Filton.

Windrush airfield (SP182119), an octagonal pillbox with a central well for a LAA gun.

Southrop airfield (SP175035), a composite pillbox comprising a rectangular section with loopholes in its longer faces and a circular, loopholed chamber at both forward-facing corners. A similar structure stood at the opposite end of the airfield, and both were accompanied by free-standing Norcon-type circular pillboxes.

Staverton airfield (SO875226), a pentagonal pillbox with an open annexe containing the entrance and a pillar for mounting an AA light machine gun. This is one of several such examples with a further one on Nut Hill.

The experimental use of dogs to guard aerodromes began at Staverton where their unerring ability to sniff out intruders led to the first Guard Dog Training School being set up in late 1941 by MAP at Woodfold in Down Hatherley, moving to RAF Staverton in 1946. It was run by Major James Baldwin, a veteran of the Great War and a respected dog-breeder. Eight dog-handlers and dogs were also trained for patrolling the extensive storage facilities of the US SOS depots at Bristol, Lydney, Gloucester, and Ashchurch. Dogs were also used to guard the SLGs at Down Farm and Barnsley Park which were experiencing problems with intruders. When the dogs were found higher priority targets to guard, armed airmen or troops took over their duties.

The Gloucestershire Home Guard

Although the Chiefs of Staff Committee retained an anti-invasion sub-group into 1942, the perceived threat of invasion was receding, and the Home Guard assumed more responsibility for national defence. In Gloucestershire there was a total of 32,000 men and women in the county's nineteen battalions. Nine of these were based in Bristol; three in Gloucester; two across the Severn in the Forest of Dean; and the remainder centred on the towns of Stroud, Cheltenham, Cirencester, Dursley, and Chipping Campden. Several of these battalions covered

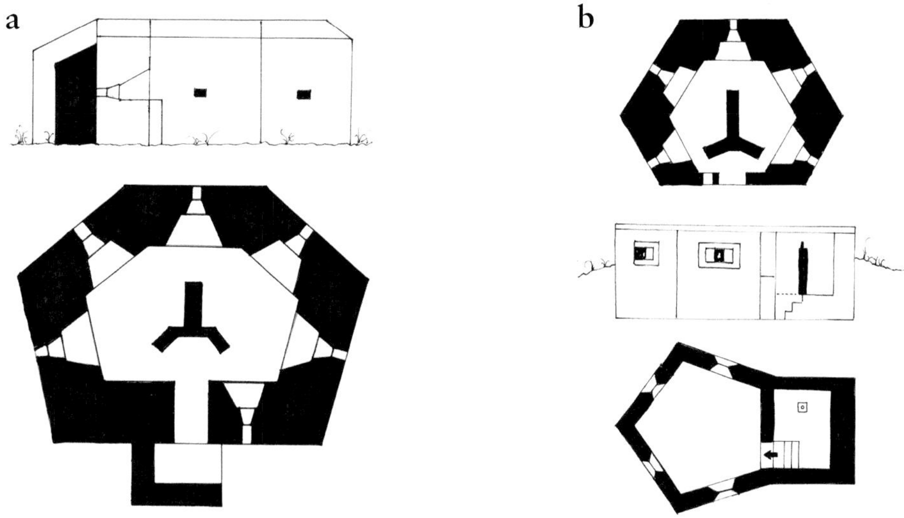

Sketch plans of pillbox types in Gloucestershire.

a) Plan and side elevation of a GHQ Green Line-type DFW3/24 shell-proof pillbox.
b) Plan of a GHQ Red Line-type DFW3/22 shell-proof pillbox with thin rear wall (*top*); plan and elevation of a locally designed pillbox with attached LAA position found only at RAF Staverton and on Nut Hill (*bottom*).
c) Plan of a splinter-proof, rectangular pillbox found on both banks of the River Severn (*top*); plan and elevation of an unusual, splinter-proof, octagonal pillbox at Lydney Harbour (*bottom*).
d) Plans of two of several Section Posts on the River Wye Stop Line (*top*); plan of an octagonal, splinter-proof pillbox with an internal corridor and a central, open LAA position at RAF Windrush (*bottom*).

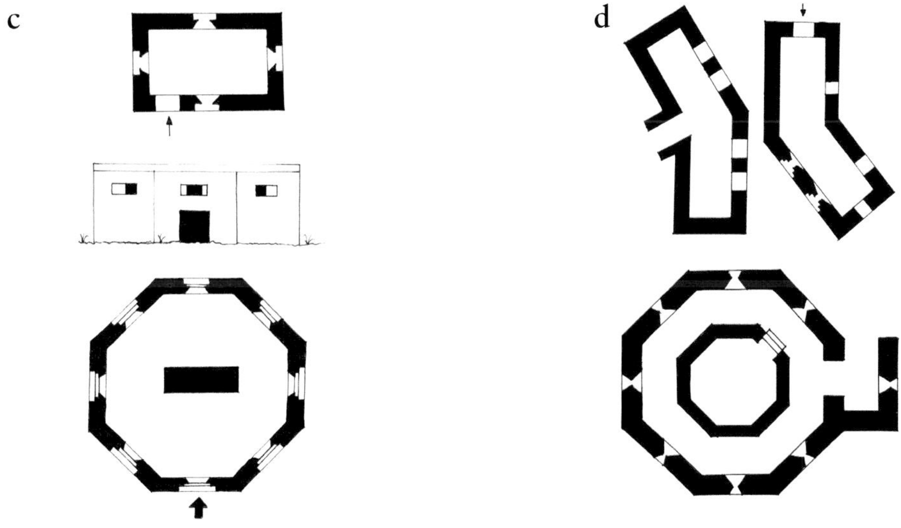

specific installations such as Filton Airport, the Brockworth aircraft factories, LMS and GWR facilities, Smith's, Dowty, Hoffman's, Sperry's, Fielding and Platt and Rotol. Many of these units were recruited from their own workforce, thereby exploiting local knowledge and established management hierarchies. In Bristol, the shipyard of Charles Hill & Sons raised a unit to man a ZAA battery defending the docks. The six troops of 71 AA Battery (Home Guard) helped to man the HAA batteries around Bristol and the LAA guns defending VPs such as those listed above. The five rocket batteries manning the ZAA were based in Bristol and Gloucester forming 17 AA Regiment (Gloucestershire Home Guard). All these AA units were affiliated to their local Home Guard battalions but trained, drilled, and paraded separately. Further specialist units were recruited to provide transport for operations and exercises. No. 2136 (Bristol) Home Guard Independent Motor Coach Platoon, consisting of twenty-six vehicles, was based at the Bristol Tramways & Carriage Co.'s Lawrence Hill depot as part of the Somerset Home Guard Transport Column; and No. 2143 (Gloucestershire) Home Guard MT Company was based in Gloucester as part of the South Midland Home Guard Transport Column.

Eden's appeal for recruits to the Local Defence Volunteers (LDV) had brought an immediate surge of volunteers. In Cirencester, 160 men enrolled, in the police station initially and then at the Old Museum. They then paraded at the militia Barracks and were given a pep talk by the C/O, outlining the course of the war. Their task would be the defence of the town with railwaymen forming a platoon to safeguard the important line from the Midlands which went through Watermoor Station, while another GPO platoon secured local communications. The newly enrolled volunteers in Cheltenham paraded in Montpelier Gardens. At the same time, the Women's Home Defence Corps (WHDC) was being formed in Bristol under the auspices of a local female MP. It went on to become a national organisation under the chairmanship of Dr Edith Summerskill, with thirty branches countrywide but never enjoying any real official recognition, respect, or support. Members wore denims with the WHDC badge and practised shooting and unarmed combat, often being trained by Home Guard members. Many women, although officially discouraged, would eventually join the Home Guard itself in a variety of non-combatant capacities.

One of the first jobs of the LDV and then the Home Guard was to guard VPs. This often involved setting up checkpoints and harassing innocent locals. The LDV commander in Chipping Campden in the summer of 1940 was the retired Brigadier Sir Percy Hobart, later to be recalled as commander of the 79th Armoured Division, charged with cracking the Atlantic Wall. His regime was so draconian that even long-term residents were disbelieved on principle and forced to prove the legitimacy of their presence. Other units in the county were no less stringent, being ordered automatically to mistrust everyone, no matter how harmless or plausible. The result of this overenthusiastic and overofficious approach would occasionally result in tragedy, as reported in the *Daily Worker* in June 1940, when an ARP warden was shot and killed in the county for allegedly ignoring a challenge. At VPs, sentry rotas were drawn up and every night, two members of the seventeen-strong Frampton Mansell platoon guarded the entrance to the Sapperton railway tunnel near Stroud. The main task of the 5th Battalion's 'E' Company was the defence of the Severn crossings north of Gloucester: the Maisemore bridge and the Ashleworth ferry. Observation posts were established at prominent points to act as an early warning system. The church tower on Chosen Hill, Churchdown, was manned by the signals section of the 19th Battalion's 'A' Company.

The major objective of anti-invasion efforts was to curb enemy freedom of movement, by creating obstacles that might impede his progress and so win time for the Field Force's arrival from its central assembly points. At Wotton-under-Edge, roadblocks were constructed out of tree trunks and cartwheels, and at Springhill near Chipping Campden, 'the Irish came and put up big poles all over the fields' to discourage airborne troops from attempting to land gliders or transport aircraft. An advert appeared in the Wiltshire and Gloucestershire press, calling for volunteers to 'assist in the construction of defensive works against possible landing of enemy aircraft'. Hawsers were stretched across roads to deter enemy despatch riders, the remnants of one such installation surviving at Down Ampney. As the hardened defences took shape, then home guardsmen would be assigned manning duties. Despite a general paucity of the means of achieving mobility, some units did what was possible. Cheltenham's dozen-strong 'Fast Cycle' platoon patrolled areas where enemy paratroops might be dropped. Their orders were to make a nuisance of themselves in disrupting the enemy's communications. Depending on the size of catchment, units would be of vastly different strengths. The Cirencester battalion's pretty representative Kempsford platoon was photographed in July 1943. There are over forty men present, including a colonel and a major, both presumably of First World War vintage, a captain and a vicar. The group is flanked by two motorcyclists and, inevitably, adorned by two dogs.

Local defence plans were formulated, and appropriate duties assigned. At Stonehouse, 'E' Company, 8th (Dursley) Battalion, had its HQ in Stanley Mill. One platoon was composed of Air Ministry employees of the Meteorological Office, evacuated to Wycliffe College, and the other of local volunteers. Their task was to defend the keep—the last-ditch, fall-back position. The Stonehouse and AM platoons would engage in highly competitive mock battles with 'H' Company, composed of platoons drawn from the employees of the Sperry Gyroscope factory at Bond's Mill and from Hoffmans' Ball-bearings works at Oldends Lane. Hived off from the 13th Battalion, the sole role of the 18th Battalion was to defend Filton Airfield. Its HQ was Fairlawn House and its Battle HQ was in a basement beneath the airfield's security office. Two of its five companies were drawn from the BAC works, while other companies were responsible for bomb disposal, firefighting and policing. Particularly vulnerable concerns were assigned entire Home Guard companies. Rotol and Glosters at Brockworth were the responsibility of two companies of the 19th Battalion; two companies of the 1st Battalion guarded Dowty Equipment Ltd. at Staverton and Smith's at Bishops Cleeve; while the 6th Battalion's 'C' Company looked after Yate aerodrome and the 10th Battalion's 'A' Company minded Whitchurch Airport. The 1st and 16th Battalions located their HQs, respectively, in the TA drill halls in Gloucester and Bristol; the 8th Battalion's was in Dursley Court; the 1st Battalion's in Lypiatt Terrace, Cheltenham; and the 14th Battalion's in Avonmouth Docks.

Home Guard units spent much time in training, using whatever premises were available. The long room of Berkeley's Swan Inn in the High Street housed lectures, as did the parish hall at Hartpury. The Cheltenham GWR platoon used the North Street drill hall and then the Ladies' College sports field. Woodchester Park in Stroud provided space for wider ranging exercises and the Cheltenham battalion held weekend camps at Seven Springs near Coberley. Specialist training was carried out at places such as the Eastwood Park Anti-Gas School in Falfield. Small-scale exercises often happened by arrangement between local units, providing ample opportunity for all sorts of underhand machinations. Rodborough Fort, built in 1764 outside Stroud as a folly with cannon mounted, was taken over by an army AA battery. One morning, the local Home Guard commander, a milkman, captured the place by concealing his men in his milk float. The regulars, recovering from a heavy night playing a darts match against a Home Guard team, were slow to awaken that morning.

As the Home Guard grew in confidence, official exercises became more ambitious. One such large scale exercise was the Second Battle of Tewkesbury, fought in May 1941. Occupying the Red Lane Drill Hall was 'A' Coy, 1st (Cheltenham) Battalion, from which had been hived off a new 'E' Company with its HQ in the Old Hospital. They would be the defenders in an exercise designed to test the defences of the bridges over the Severn, and the efficiency of the liaison and communications systems between the different units across county borders. Also part of the defending force were RASC troops from Ashchurch plus two companies of the Gloucester battalion and Home Guard units from Worcestershire. The enemy force was represented by troops of 37 S/L Regiment stationed at Mitton to the north of Tewkesbury and others drawn from 162nd Infantry Brigade, consisting of first-line TA units, advancing from South Wales, and aiming for the industrial heartlands of the West Midlands. Umpires from the Regular Army, based in Cheltenham's Dean Close School, observed outcomes and raised issues for exploration and incorporation into subsequent training sessions. Joint exercises were held in Bristol in 1942 with the field army fulfilling the role of attacker and the defence entrusted to the Garrison and the Home Guard.

As the war progressed, the Home Guard armoury increased in size and efficacy, attempting to increase its numbers of automatic weapons. As early as 1940, the reformed Cheltenham GS cadet corps supplied a Lewis gun section to the local Home Guard battalion. Photographs of the Crickley Hill platoon of the 19th Battalion's 'A' Coy, show men equipped with Browning Automatic Rifles. Alongside the conventional small arms and regulation hand grenades, a whole arsenal of alternative improvisations evolved. In the early days, sticky bombs were issued to Home Guard units, along with mines on casters for pushing under tanks which had been stalled by obstacles. The lack of proper artillery saw the development of 'sub-artillery' such as the Blacker Bombard or spigot mortar, the Smith gun, and the Northover Projector. This last weapon was a drainpipe on legs from which were projected Molotov Cocktails, officially, Self-Igniting Phosphorous (SIP) bombs. These weapons were often more perilous to the user and could be dangerously inconsistent and the Painswick Platoon practising on their Northover Projector could attain no better than a 50 per cent success rate. Produced by a local foundry, the Cirencester Home Guard ran an armoured vehicle. This was a Dodge truck chassis with steel plate on the front and a fighting compartment to the rear formed from sand-filled boxes with firing loops. By 1942, Home Guard signals sections were equipped with the No. 38 field radio, but pigeons were still retained as message carriers of last resort.

Somewhat surprisingly, the Home Guard also fulfilled an important ceremonial role. Parades such as those held by the GWR in Gloucester served to attract recruits and to demonstrate the developing professionalism of the Home Guard. In Cheltenham, the GWR platoon provided the guard of honour for royal visitors arriving by train. A 'Wings for Victory' fundraising parade in Hartpury, led by the 5th Battalion's 'E' Company in May 1944, shows a variety of weapons being carried. When King Haakon of Norway visited the Filton aircraft works in 1942, the guard of honour was provided by the resident 13th Battalion. The Home Guard's final parades were held at the end of 1944 when the force was stood down. Queen Mary, who had spent the war at Badminton House, attended the 5th Battalion's thanksgiving service in Gloucester Cathedral, which was followed, a week later, by the stand-down parade in Gloucester Park.

The Air War in Gloucestershire and Bristol

As we have seen in the previous chapter, Gloucestershire's major contributions to the air war would consist of production, training, and logistics: aircraft storage; and assembly, supply, administration, meteorology, and maintenance. Most of the flying which took place was thus confined to training, but Filton's Spitfires and Hurricanes were active up until 1941, and Hurricanes flew from Bibury in the last four months of 1940. The OTUs at Moreton-in-Marsh, Aston Down, and Chedworth trained crews through active participation in operational sorties, and in 1944, gliders from Down Ampney and Fairford took part in operations over northern Europe.

Air Operations

Filton was listed by the Air Ministry as a bomber station in 1930 and in 1936, but by March 1939, it was part of Fighter Command, and it was Hurricanes of 504 Squadron which broke up German bombing raids in September 1940. Operational fighter squadrons were stationed at Filton until August 1941 when the barrage balloon defences of the BAC works made flying even more hazardous than normal. In 1939, Aston Down was home to No. 12 Group Pool training Spitfire pilots, and then No. 5 OTU, whose pilots flew Hurricanes by day or Blenheims by night, often engaging, and sometimes shooting down, enemy aircraft. In the autumn of 1940, Hurricanes of Nos 87 and 92 Squadrons flew defensive sorties out of Bibury claiming enemy aircraft. Early in 1941, the training of Wellington crews of No. 21 OTU at Moreton-in-Marsh involved leaflet-dropping sorties over Paris; bombing raids on the Pas de Calais; diversion flights over the North Sea to mislead German night-fighter controllers; and more leaflet-dropping and bombing raids prior to D-Day. In 1945, Typhoon fighter-bombers were based at Chedworth for a short while.

A major element of Gloucestershire airfields' training and operations related to glider-borne operations. Fairford's initial complement was fifty Short Stirlings and 100 Horsa gliders. Stoke Orchard and Long Newnton were Glider Training Schools and Fairford was a heavy glider conversion unit (HGCU) for introducing glider pilots to the larger Hamilcar gliders, some of which could carry the Tetrarch light tank. Part of the training at Fairford, which had predominantly been focused on glider-borne landing operations, also allowed some crews to fly missions to drop supplies to the French Resistance under the auspices of SOE. The gliders would be flown by personnel of the Glider Pilot Regiment belonging to the four flights of 'G' Squadron. The Stirlings also practised dropping paratroops. On D-Day, forty-five Stirlings dropped nearly 900 paratroops of the 6th Airborne Division close to the Orne bridges, and then did a quick turnaround so that thirty-six of them were able to take part in Operation Mallard. This involved delivering jeeps and trailers, 75-mm guns, motorcycles, and 245 men to the Normandy battlefield, one glider-load of which landed slap-bang next to Pegasus Bridge. Confined to a secure 'cage' at Fairford, prior to embarking on their operations, the best known of which was Bulbasket, SAS units were dropped into France. The three elements of Operation Market to capture the Arnhem bridge over the Rhine were disastrous for the Stirling squadrons. Almost 100 successful sorties, delivering gliders with troops and equipment, were followed by a further fifty sorties to drop supplies, but sixteen of the fifty aircraft in total involved were lost to enemy fire. Gliders operating from Down Ampney were involved in Operations Tonga, Mallard, and Market 1&2 using Horsas towed by Dakotas. After Market, as the operational squadrons were moved nearer to the continent, Down Ampney and Fairford, as No. 22 HGCU, resumed a training role in an attempt to rebuild the shattered squadrons of the Glider Pilots Regiment.

The airfield at Yate continued to be used throughout the war by Hendy Hecks carrying out liaison work. The US Army also employed light aircraft for liaison and courier duties linking far-flung units. Beggar's Bush Field, which had constituted Clifton College's playing fields, became an airfield for the light aircraft which provided some of these links. Lilleybrook golf course at Cheltenham was used by US SOS liaison aircraft. Whitchurch and Filton were bases for the civilian men and women who flew unarmed aircraft as ferry pilots for the Air Transport Auxiliary. Although there was a proposal in early 1944 to deploy a USAAF IX Air Force light bomber wing to Fairford and Down Ampney, nothing would come of it. Similarly, speculative plans for USAAF Carrier Command airfields at Barnsley Wold, Tresham, and Shipton Moyne were not followed through. Chedworth was used for a few weeks in the summer of 1944 by the IX USAAF 125th Liaison and HQ Squadrons flying L-5 Sentinels. The US VIII AAF maintained a truck depot in Bristol from 1943 which was passed on to the US Tactical Air Force.

Aviation Training

The recruitment drive begun in 1938 continued throughout the war but it became increasingly difficult for youngsters to join the RAF in more or less any capacity without a good level of maths and science attainment. The Air Training Corps, a unit of which formed at Cheltenham GS in 1941, helped to bring cadets up to a satisfactory level, and youngsters who had left school at fourteen, as most had done, could attend evening classes in mathematics and engineering. Young maths and physics graduates were recruited to prepare those who stayed on at school, for Matriculation and beyond, to become potential RAF officers. A teacher from Farmor's School at Fairford was even seconded to run the RAF Canteen at Cheltenham Technical College. Innsworth opened in 1940 as No. 7 School of Technical Training for engine and airframe fitters and mechanics. No. 2 WAAF Depot opened there in December 1941 and training of WAAF personnel, particularly for barrage balloon operation, became the station's predominant activity, while also housing a number of other administrative functions mainly relating to personnel issues.

In September 1939, the HQ of RAF No. 23 (Training) Group moved into South Cerney. All levels of flying training from basic through intermediate to advanced were carried out at South Cerney, Little Rissington, Stoke Orchard, and Staverton. Specialist courses were also mounted as a standard blind approach was installed at Babdown Farm in 1943; other blind approach training took place at South Cerney; and Staverton was home to an air observer navigation school. In 1942, Flight Refuelling Ltd moved into Staverton to carry out experimental work, but only in 1944 were they contracted to equip 600 Lancasters as tankers and to train the 600 Receivers necessary for their operation. Babdown Farm was a flying instructor's school, 1942–43; Moreton Valence was a pilot refresher flying training school in 1942; Aston Down and Chedworth both hosted the fighter leaders' school, during 1943; Harvard trainers from Chipping Norton flew into Windrush for night-flying practice; and by 1945 Little Rissington had been adopted as the RAF Central Flying School. Babdown Farm, Southrop, and Northleach were all relief landing grounds (RLG) for flying training schools and Oldbury Sands was a bombing range for use by OTUs.

RAF and USAAF Airfields and Associated Structures

In contrast to the ordered elegance of the expansion period stations, the new wartime builds relied on the principle of dispersal. The planting of these sprawling airfields on the landscape could create quite an impact on rural settlements. At Down Ampney, the village school's playing field was taken over and the children of some families were forced to change school as the airfield had sealed off the access by public road. Technical, administrative, domestic, or communal sites were all separated out, and aircraft would be similarly dispersed on hard standings scattered around the airfield perimeter. Most of Gloucestershire's airfields were grass, with tarmac or concrete runways appearing only slowly. This made them particularly susceptible to weather related problems. Units commuted between Northleach and Stoke Orchard as both fields were apt to break up in bad weather. Moreton-in-Marsh was built by Wimpey in 1940 with three tarmac runways but they were judged as being too short and had to be extended. Various designs of pierced steel planking were used to reinforce grass strips. A Sommerfeld track runway was laid at South Cerney, two at Bibury and Long Newnton, and three more at Babdown Farm, while the older 'Army Track' was laid at Southrop. Chedworth was laid out to the 'Crossbow' design with just two runways crossing at right-angles to each other. Windrush, too, had only two runways but in the shape of a St Andrew's cross. At the Glosters factory at Brockworth, a single runway was laid in late 1941 to serve the two factories building Hurricanes. Filton, which had started the war as a grass field, was provided with two concrete runways, respectively 4,000 and 4,500 feet (1,230 and 1,385 metres) in length. Fairford was intended for the USAAF, and construction was begun in 1943 by McAlpine's, but on completion it was reallocated to the RAF. There was a 6,000-foot-long (1,850-metre-long) main runway, fifty dispersals, and a pair of T2 hangars. By 1943, standard blind approach had been installed at Whitchurch and Moreton-in-Marsh, with Bibury, Little Rissington, and Babdown Farm becoming similarly equipped by the end of 1944.

Wartime airfield buildings were generally of prefabricated design using easily available materials and simple construction techniques. Moreton-in-Marsh would be the last station to get a pre-war J-type hangar as subsequent hangars were either Bellman or T (Transportable) types consisting of easily erected steel frameworks clad with corrugated-iron sheets. Bibury, Long Newnton, and Windrush each had a single T1, while Babdown Farm had three; Down Ampney and Fairford each had a pair of T2s and Moreton-in-Marsh had four; Stoke Orchard had four Bellmans and Staverton had at least one T2, now heightened and modernised, but often referred

Long Newnton airfield, the watch office built to a 1941 design for Bomber Satellite or Operational Training Units.

Stoke Orchard airfield, one of its four Bellman hangars with a Marston Shed behind.

Babdown Farm airfield, one of the airfield's three T1 hangars.

Windrush airfield, a Blister hangar.

to as a Bellman, along with smaller, pre-war sheds. MAP designed the B1 hangar as an aircraft repair shed and erected one at Moreton Valence for the development work being carried out there by Glosters. When the USAAF moved into Filton they erected four portable Butler hangars. Blister hangars were introduced in the early days of Babdown Farm and thereafter in large numbers: five at Bibury, seven at Southrop, a pair at Chedworth, and eight at Windrush.

Where brick was used, it was normally for walls of single thickness with external buttresses, a style known as 'temporary brick' (tb). The last peacetime design of watch office (518/40 plus 8936/40) was built at Moreton-in-Marsh but a range of specialist watch offices was then initially designed for use at designated airfields, including the 13726/41 for Satellite OTUs at Bibury, Filton, Southrop, and Windrush. These were then superseded by a general-purpose watch office, designed for those airfields which warranted their presence such as Aston Down which had the prototype 12779/41, and the final 343/43 design at Moreton Valence, Babdown Farm, Down Ampney, and Fairford. Most of the rest of the airfields' myriad buildings were either tb or timber huts or the various other ranges of hutting, the most common being Nissens and their big brothers, the Romneys, used primarily for main stores and workshops. Laing huts were used at Bibury. More of these lesser, temporary buildings may be seen, particularly, at Chedworth and Long Newnton.

The Aircraft Industry

Despite the conviction that the county would remain out of reach of enemy bombers, serious raids soon affected production. A raid on the British Aeroplane Company (BAC) at Filton by fifty-eight Heinkel bombers on 25 September 1940, resulted in the deaths of 132 people, ninety-one of whom were aircraft workers, with over 300 seriously injured. A direct hit on the tool room's earthen air raid shelter killed forty-five skilled toolmakers, an extremely grievous loss to the company's manufacturing operation and local families. In November 1940, and again the following March, the Douglas/Aero Engines works at Kingswood was hit by incendiaries and HE bombs causing casualties and minor damage. Earlier raids had caused some superficial damage to Parnall's aircraft factory at Yate, but on 27 February 1941, an attack by a lone intruder, flying at an altitude making it unseen and unreachable, caused devastation, resulting in fifty-two deaths and over 150 injuries among the 4,000 workers. The wrecked factory forced production to be immediately dispersed to other sites including Boulton Mills in Dursley. Workers were bussed all over the area and only one week's production was lost. Hucclecote aircraft factory underwent raids in March and April 1942. The solution proposed by Lord Beaverbrook at MAP was to go underground. This had followed a proposal for securing safe ammunition storage back in

1930, when Avening Quarries near Nailsworth had been a possible venue but, in the event, found to be unsuitable. In 1940, part of Eastlays Quarry between Melksham and Corsham in Wiltshire had been suggested as a site for storing BAC's Centaurus engines, but the scheme was rejected as being too expensive. However, a plan to move BAC underground, into the two million square feet of Spring Quarry near Corsham, was one of the five major initiatives promoted by Beaverbrook with Churchill's enthusiastic approval but accompanied by a distinct lack of enthusiasm from the Treasury. As time passed and the costs of converting the quarry mounted from the initial estimate of £1 million to £6 million, the need itself would evaporate. By 1942, BAC was planning to move its aero engine production into Spring Quarry along with its development works, Parnall's Frazer-Nash gun turrets and the production of undercarriages for Halifax and Lancaster bombers run by Dowty's, altogether employing up to 14,000 workers. In the midst of spiralling costs and insurmountable construction problems, the Gloucestershire-based firms' needs for alternative accommodation melted away. Spare factory space at AV Roe's other sites meant that Dowty no longer needed its allocation and Parnall's learned that they would be able to cut their production as the heavy bombers being flown in from the USA were already fitted with gun turrets. Meanwhile, BAC's move to Spring Quarry went ahead but the enormous technical problems besetting the production of Centaurus engines were only compounded. Output of the Centaurus, projected to be 260 per month, averaged out in the end at only twenty-five. Development of other experimental aero engines: both the gas turbine Theseus and the high-compression Proteus enjoyed greater success but would be completed elsewhere. Given that the Spring Quarry factory was only in operation for eighteen months at a cost of £20 million, had been unwanted by the majority of its tenants, and was redundant by 1942 when most of the bombing had ceased, then it should have been aborted. Beaverbrook unfortunately persisted in refusing to recognise the waste and was supported by Churchill who could never resist a gimmick.

Notwithstanding all these diversions, the local aircraft industry produced impressive results. Gloster's at Brockworth built 2,750 Hurricanes in 1939–42 and 3,300 Typhoons 1941–45. Next door, their Hawksley subsidiary at Hucclecote, a result of the amalgamations of the 1930s, built over 600 Armstrong Whitworth Albemarles. The Hucclecote assembly shop was 450 yards (415 metres) long and the Brockworth shadow factory covered 24 acres (10 ha). In August 1940, Short Brothers' factory at Rochester in Kent was bombed, causing serious damage to the Stirling bomber production line. As a stopgap, the damaged aircraft, with their operatives, were hurriedly moved to Glosters at Brockworth, some twenty-eight of them being completed before the new Short's dispersed facility at South Marston, Swindon, was completed. Shorts also had a dispersed facility in Gloucester as JH Lower, and Stirling fuselages were fitted out at Friars Hill, Sevenhampton. Glosters were contracted to design and build an airframe for Whittle's new turbo-jet engine in 1940. This was to become the first Allied jet aircraft, the E28/39, two prototypes of which were built, one at Brockworth and one at the Regent Garage in Cheltenham where it was guarded by a single armed police officer. Its inaugural taxi took place in April 1941 when Gerry Sayer, Glosters' chief test pilot fired the engine up at Brockworth. The first real flight-testing—a series of fifteen flights over twelve days—staged at Cranwell in Lincolnshire, scene of Whittle's apprenticeship training, showed the potential of the new aero engine technology, with Sayer reaching 370 mph (592 kmh) in what Whittle and his team referred to as the 'Squirt'. Apart from Hucclecote, Glosters had several other shadow factories in the area. Bentham was employing 130 workers in 1942 who would begin production of the twin-engined F9/40 Meteor jet fighter the next year, with test flights being made from Moreton Valence. They also occupied a piano factory in Woodchester, and had a shadow factory, to the north of Stoke Orchard airfield, assembling Typhoons. Although Carter envisaged the development at Glosters of heavy bombers powered by jet engines, for MAP, still lukewarm over Whittle's inventions, it was a step too far.

BAC at Filton with its 32,000 employees working in 2.6 million square feet (25 ha) of factory space, and its HQ at the Royal West of England Academy at Clifton, produced Beaufighter multi-role aircraft. The Bristol Hercules aero-engine, nearly 60,000 of which were produced, powered many of the aircraft then in production, and towards the end of the war, over 100,000 Centaurus radial engines were built. Rover Cars also manufactured Mercury and Pegasus engines for BAC at its shadow factories in Solihull and, underground, at Drakelow. Almost 6,000 Beaufighters were built at Filton; at BAC's two shadow factories: Weston-super-Mare operated by BAC and the other by Fairey Aviation in Stockport; with more built in Australia. Dowty Equipment Ltd made parts for twenty-eight different aircraft including the Hurricane, the Lancaster, and most of the aircraft produced by BAC and Glosters. Their major products included nearly 90,000 landing gear assemblies and almost one million hydraulic units, some of the work being carried out by subsidiaries in Canada and the USA.

Parnall's produced their gun turrets and became one of the largest manufacturers of Spitfire components, rebuilding their Yate factory in Station Road by 1944. Other manufacturers were also involved in aircraft production: John Hall & Sons of Bristol made paint for aircraft; New Insulation Co. Ltd of Bristol Road, Gloucester, made wood veneer and resin, marketed as Permali, for propeller blades; Williams & James, of Lower Barton Street, Gloucester, made air compressors; HH Martyn (Aircraft) Ltd built a new factory at Sunningend, Cheltenham, in 1941, now the Lansdown Industrial Estate, where parts were manufactured for the De Havilland Mosquito. Aero Engines Ltd of Kingsdown, formerly known for their Douglas motorcycles, received contracts from MAP for wings and other parts for Supermarine Spitfires, Blackburn Bothas, Boulton-Paul Defiants, and Fairey Fireflies as well as for generator sets. Drawing on their

Cheltenham, Sunningend aircraft factory. (*Pam Osborne*)

previous experience of aero engines, the company collaborated with Rolls-Royce and others to develop new models. Production of electric trucks, motorcycles, and internal combustion engines continued throughout the war, and a partnership with Philco brought the company into the telecommunications market with bright prospects for the period of post-war adjustment. The Pine End Works at Lydney Harbour manufactured wooden parts for gliders and other aircraft and Fielding and Platt in Gloucester produced presses for BAC and Fairey Aviation along with other machinery used for making laminated wooden propellers and in the manufacture Spitfires and Hurricanes.

A far more effective solution to munitions factories' vulnerability to air attack without the high costs of going underground, was simple dispersal. Shadow factories included Hoffman Manufacturing making ball- and roller-bearings at Oldends Lane, and Sperry Gyroscopes at Bond's Mill, both in Stonehouse. Ball-bearings were absolutely vital as every Lancaster bomber needed 950 of them. Rotol airscrews built a shadow factory at Staverton, using the adjacent airfield for flight testing. The Air Ministry was concerned over the vulnerability to bombing of the important Smith's Instruments company at their works in the Waterloo Road and at Cricklewood in north London. A dispersal factory was therefore planned and a site at Kayte Farm, Bishop's Cleeve, to the north of Cheltenham was selected, allegedly because of its proximity to a prominent racecourse, but basically for its situation on the notional Bristol-Falkirk line, which was thought, at the time, to mark the western extent of the range of German bombers.

Production of aircraft instruments would overtake that supplying the automotive industry which itself would soon turn to manufacturing aircraft. Cleeve Grange was acquired to house recently purchased machinery for the manufacture of hairsprings and the first element of the new factory opened in May 1940. Smith's war production also extended to spark plugs for Rolls-Royce Merlin engines.

Aircraft Storage and Supply

Prior to delivery to operational airfields, the output of the aircraft factories was stored at Aircraft Storage Units (ASU), two of the largest being Aston Down and Little Rissington, both equipped with large numbers of hangars for storage and maintenance. Such ASUs stored thousands of different aircraft, peaking across the country at 28,000 in 1942. One of Aston Down's hangars was run by Vickers as an aircraft repair shop for Wellingtons. No. 8 MU at Little Rissington held thirteen different models of aircraft in store in 1940, and sixteen in 1941, before the system was rationalised so that, by 1944, each ASU held only five or six types. At Aston Down, No. 20 MU was relegated to the role of lodger but continued to operate throughout the war.

A complement to the ASU was the Satellite Landing Ground (SLG) which could hold between fifty and a hundred aircraft, usually out in the open and often under tree cover. A provisional list was produced by Sir Alan Cobham in 1941 which included Macaroni Down, near Eastleach, Overley near Cirencester, and Calmsden near North Cerney. A secondary list cited Kingscote near Nailsworth and Cirencester Park as possible SLGs. In the event, Calmsden was not used, and Macaroni Down was made ready but then de-requisitioned, only three sites in the county becoming operational. Overley, No. 14 SLG, was operated by No. 20 MU at Aston Down. It had a standard bungalow HQ/guardroom and a Robins hangar supplied by RAF Kemble in 1943. Although provided with a large expanse of Square Mesh tracking, it appears not to have stored great numbers of aircraft but was nevertheless regarded as a major site. Barnsley Park, No. 22 SLG, was operated for only a short while by No. 6 MU at RAF Brize Norton (Oxfordshire). Even after extensive improvements, it remained very difficult to operate, only, as a satellite of No. 5 MU at Kemble, becoming busy in 1944. A bungalow HQ building, now converted for domestic use, lies to the west of the single airstrip along with a Super Robins hangar and equipment huts. Aircraft storage and dispersal areas lay both to the north and south of Barnsley Park Farm. Down Farm, No. 23 SLG near Tetbury, run by No. 10 MU, was associated with Hullavington (Wiltshire). The Air Ministry had requisitioned both the nearby Westonbirt School and the Arboretum in 1939 and, after 1942, stored aircraft in the grounds of both. The SLG itself, with resilient grass, Sommerfeld Track hard-standings installed in late 1943, and a ready water supply from the school's reservoir, was capable of being used for the storage of four-engined aircraft, particularly Stirlings used as heavy glider tows. Its two airstrips, forming a T-shape, were also used for circuit training by aircraft from a waterlogged Babdown Farm. From early 1944 until the end of the war, there were never fewer than 100 aircraft on site. Little Rissington's SLGs at different times were No. 3 SLG, Middle Farm and No. 28 SLG, Barton Abbey, both in Oxfordshire.

Gloucester was home to a Purgatory Unit where incomplete aircraft were stored temporarily. When airframe production overtook that of engines or propellers, congestion in factories was relieved by borrowing engines or props so that the completed airframes might be delivered into storage. The borrowed or 'slave' parts were then returned for installation in another airframe. A shortfall of Rotol Electric Propellers in 1941 meant that Wellingtons were often delivered into such purgatory areas. Filton became an Overseas Aircraft Preparation Unit, receiving incoming aircraft for readying prior to issue. From 1942, Filton became USAAF Station 803 for three squadrons of the IXth Air Force Tactical Air Command. As No. IX Base Aircraft Assembly Depot, they assembled cased aircraft that came into Avonmouth Docks; recovered and repaired damaged aircraft; and serviced and maintained operational ones.

Aircraft were delivered by male and female pilots flying for the Air Transport Auxiliary (ATA) and based at both Filton and Aston Down. In 1942, a Ferry Training Flight was established at Moreton-in-Marsh flying Wellingtons to the Middle East. The next year, this flight evolved into No. 311 Ferry Training Unit, continuing to deliver Wellingtons for a further twelve months. An exploration, based at Staverton, into the possibility of those Wellingtons towing Spitfires to Malta, perhaps fortunately, never came to fruition. Before the war, the RAF had established a contract with the US manufacturer Lockheed, with an initial order for 200, to supply a military version of the Electra passenger aircraft. This was the Hudson light bomber/coastal reconnaissance aircraft, of which almost 3,000 would ultimately be built with over half of them being supplied to the RAF. To preserve US neutrality, from the autumn of 1940, Hudsons were towed over the Canadian border and flown from Newfoundland to Aldergrove in Northern Ireland. Owing to its proximity to the Air Ministry's long-range W/T and D/F station at Birdlip, Overseas Aircraft Movement Control Unit (OAMCU) run by RAF Maintenance Command but with strong links to MAP, was set up at their Regional Control Centre on Eastern Avenue, Barnwood, Gloucester. When transatlantic flight control was transferred to Prestwick in 1941, under the newly formed RAF Ferry Command, the Gloucester control centre, as Overseas Air Control (OAC) took on the oversight of flights to Gibraltar, West Africa, and the Middle East. From 1944, the Flying Control Centre was one of several supplying military aircraft with D/F, meteorological and diversion information. Whitchurch aerodrome had been requisitioned in 1939 by the Air Ministry's Civil Aviation Department, and hard taxiways and a 3,000-yard-long runway were laid, and five hangars were built. Civil flights continued from there to Lisbon and to Foynes in Ireland where passengers transferred to transatlantic flying-boat services which took them to the USA via the Azores. It suited all parties to maintain a route in and out of Britain and flights were generally undisturbed, so the shooting down of Leslie Howard's aircraft in 1943 by German fighters was unexpected.

Pucklechurch, one of the two remaining hangars for barrage balloons.

The Air Defence of Gloucestershire and Bristol

Bristol and Gloucester were designated as Gun-Defended Areas (GDA) meaning that their AA defences were co-ordinated as part of the Air Defence of Great Britain (ADGB) system, integrating guns, searchlights, the observer network, radar, balloon barrage, and fighter aircraft. Recalled from the south coast to meet the German bombing offensive in the late autumn of 1940, Nos 501 and 504 Fighter Squadrons, RAF, alternated at Filton but when they were needed elsewhere, their Hurricanes were temporarily exchanged for Gloster Gladiators. From August 1940, it had been agreed that fighters did not mix well with barrage balloons and no more operational flying would take place from Filton. The fighters moved to Bibury, and Colerne in Wiltshire assumed the role of Sector Airfield taking responsibility for the protection of the Severn ports. From 1941, Bristol came under No. 8 and Gloucester under No. 9 AA Division RA. Pucklechurch was No. 11 RAF Balloon Centre, a significant proportion of its staff provided by the WAAF. Only with the development of GCI radar stations, working in conjunction with the new Bristol Beaufighter night fighter, would Bristol be assured a more reliable level of protection.

Anti-aircraft Defences

The two HAA regiments of 46 AA Brigade, 76 at Bristol, and 98 at Cheltenham and Moreton-in-Marsh, along with 66 S/L Regiment at Horfield Barracks were responsible for the two GDAs, with particular emphasis on the various aircraft factories, especially Filton and Brockworth. From September 1939, 76 HAA Regiment's eight 3.7-inch guns were deployed to defend Avonmouth Docks, 236 Battery occupying the old Markham militia camp at Pill, but once it had become apparent that Filton airfield was relatively unprotected, they were hurriedly moved there. Personnel also defended Parnalls at Yate, manning sixteen Lewis guns, until relieved by 68 S/L Regiment, while 98 (Gloucestershire) HAA Regiment was despatched to Manchester. In the absence of specialist LAA troops, the 1st Battalion, Monmouthshire Regiment, having been converted to 68 S/L Regiment in 1938, had taken over the manning of Lewis guns at Avonmouth Docks, Parnall's factory in Yate, and Rotol Airscrews at Staverton. One company with its HQ in Tetbury was even moved on as no equipment was available. The regiment was then stationed in Thornbury, initially based in the Exchange Hotel Assembly Rooms, manning positions at Aust, Marlwood, Earthcott, Almondsbury, Rockhampton, Baden Hill, and North Nibley, later to become its HQ. Despite constant reorganisation of AA units, the regiment remained in the Bristol and Gloucester GDAs until 1944. Attached to 9 AA Division, 37 S/L Regiment, which had been evacuated from Brittany in May 1940, was charged with defending the Gloucester area.

From those very weak beginnings, the Bristol GDA would develop a formidable defence against aerial attack. By July 1940, there were thirty-six HAA guns defending Bristol, under the direction of the gun operations room in Worrall Road, Clifton. A total of twenty HAA sites, seven of them manned by 76 HAA Regiment, were constructed, armed with a mixture of static and mobile 3.7-inch HAA guns, most of them equipped with gun-laying (GL) radar. By 1943, when the worst of the bombing was over, all but six of the sites remained fully armed, four of them having been upgraded to 4.5-inch guns, and five sites, including Purdown, having gained emplacements for two extra 3.7-inch guns. In January 1943, the provision of guns had risen to a total of forty-four 3.7-inch guns in eleven batteries and a further sixteen 4.5-inch guns in four batteries, all but three of those fifteen batteries being equipped with GL radar. Gloucester GDA had eighteen sites, of which just five were ever armed, and only two of those equipped with GL radar. The shortfall in LAA provision would not really be addressed until well into 1941. Filton was provided with seventeen sites each equipped with either one 40-mm Bofors LAA gun or three Lewis guns and by 1943, there were eight Bofors guns on four sites. Yate, Brockworth, Stonehouse, and Cheltenham's three discrete VPs (SOS, Dowty, and Rotol at Staverton), enjoyed a comparable level of provision, each with at least four Bofors equipments and their accompanying GL radars.

Both AA and S/L units were constantly on the move. For S/L units, this was partly due to changes in tactics. At first, searchlights were deployed in lines about 3 miles (5 km) apart, forming grids. They were then brought together in clusters of three, spaced 5 miles (8 km) apart. At Filton, four such clusters from 354 Battery, 39 (Lancashire Fusiliers) S/L Regiment, were placed close together around the airfield to experiment in Dazzle Defence which was designed to blind incoming aircraft. These techniques appeared not to be getting results and a new configuration of discrete zones for searchlights, now adapted to carry their own control radar; for guns with GL radar; and for fighters carrying AI radar, would eventually be established, with searchlights operating in Indicator or Killer belts. As communications improved, specialist signals units took over the running of gun operations rooms, that in Gloucester being operated by a mixed signals company which included ATS personnel. By 1942, Nos 8 and 9 AA Divisions had been merged into No. 3 AA Group, aligned with No. 10 RAF Fighter Group, and covering the whole of Southwest England and the southern half of Wales, with RAF Colerne established as the local Fighter Command Sector Station.

The county's forty-odd HAA batteries were all built to a standard plan with an arc of four emplacements for 3.7-inch guns behind which stood a semi-sunken command post and two magazines. The emplacement was octagonal with low walls backed by ready-use ammunition lockers. In the centre was the holdfast for the gun and later on, when guns were powered electrically, channels for the wiring. Five of Bristol's sites were given two extra guns to be mounted in simple square emplacements. A camp of huts, workshops, garages, messes, and offices stood a little way from the guns. Several of these HAA sites have survived with that at Purdown being the best preserved and most easily accessible, one of its square and all four octagonal emplacements survive along with the command post and magazines. The camp lay where the microwave tower now stands and

Filton airfield, a tower for mounting a 40-mm Bofors LAA gun with added superstructure but now long destroyed.

Purdown, the command post for the HAA site.

Tetbury, the mobilisation store for AA guns and equipment.

at the bottom of the hill. Less impressive remains survive at several other sites, including Cribbs and Rockingham Farm, but others like Almondsbury, buried under the M4/M5 interchange, have totally disappeared. LAA sites, on the other hand, were generally much simpler with a holdfast for a Bofors gun or a slot embedded in a concrete pit into which could be inserted the stalk (one design was actually called a 'Stork' mounting) for a pair of machine guns or cannon. Particularly at factories, these were often mounted on the roof of a convenient building. The Hoffmans' HG platoon in Stonehouse was equipped with four 20-mm LAA guns and twin Browning machine guns with a built-in seat, mounted on a revolving pivot. To gain height, some LAA guns were mounted on towers, either of solid brick or a steel framework. At the north end of Filton airfield (ST604812) were brick and concrete twin towers mounting a single Bofors gun on one and its accompanying instruments on the other. A similar structure stood at Portishead Docks. Searchlight sites, by contrast, have left barely perceptible remains, often no more than a shallow bowl often mistaken for a prehistoric disc barrow. One S/L site at Fretherne was manned by No. 349 Battery, 37 Regiment, and several of the regiment's other sites can be identified in post-war aerial photographs. These include Ashchurch Depot, Boddington, Churchdown, Bisley-with-Lypiatt, Mitton, Rodborough, Miserden, and Painswick. Their searchlights were operational during the Baedeker raid on Cheltenham of 27 July 1942.

As experienced AA units were moved out of AA Command to join the Field Force, men from the Home Guard and women from the ATS were deployed to mixed batteries. A new unit, 133 HAA(M) Regiment, was set up at Clifton and took over the manning of eight of Bristol's batteries including Portbury, Cribbs, Rockingham Farm, and Westbury. Members of 'A' Company of the Tewkesbury Home Guard had already been employed as shell handlers for the 3.7-inch HAA guns at Staverton airfield being bussed in by army trucks each day, but Home Guard involvement soon expanded. Four of Bristol's HAA sites, including Reservoir, Whitchurch, and Purdown had been handed over to 71 (Gloucestershire) HAA Home Guard Battery by 1943. There were already five LAA batteries, drawn from the 15th and 18th Battalions, defending Filton airfield and the BAC factory, with further batteries defending Glosters at Brockworth/Hucclecote, Rotol at Staverton, Hoffmans at Stonehouse, and Smiths and Dowty at Cheltenham. There was also work for the Home Guard manning the ZAA rocket batteries. Each Z battery deployed sixty-four Twin Projectors launching 3-inch Unrotated Projectiles (UP) in a single salvo of 128 rockets. Each of these rockets had the same explosive impact as a 3.7-inch AA shell and the intention was not necessarily to hit individual enemy aircraft but to distract, unnerve and scatter their formations. Collectively, the six ZAA sites provided the nearest thing to an AA barrage that would ever be achieved. Five of these sites were in the Bristol area and the sixth was split between Gloucester's dog track and at Staverton airfield. Whatever their benefits, they were balanced by being labour intensive. Needing at least 150 personnel on each site every night it was necessary to involve large numbers of the Home Guard in their manning. The new 17 AA Regiment, Gloucestershire Home Guard, consisted of six troops designated as rocket batteries. Since all their personnel had day jobs, they could only be expected to turn out one night in five and this necessitated huge overmanning. Training took place in Horfield, at the Wellington Garage and the RASC Garage at the Drill Hall.

Deception: Bombing Decoys, Camouflage, and Beam-bending

The air defence armoury contained other weapons mainly based on deception. One way of diverting bombers from their intended targets was to provide alternatives. These bombing decoys were provided initially for airfields. Daylight, or 'K' decoys, consisted of dummy airfield installations and aircraft, produced by technicians from the major film studios, which were moved around by ground crew to simulate realistic activity for the benefit of *Luftwaffe* reconnaissance missions. Night-time 'Q' decoys were dummy flarepaths lit from a manned control bunker and extinguished just after the sound of approaching enemy aircraft had been heard. Sometimes decoy fires (QF) were lit to simulate marker bombs dropped by pathfinder aircraft. Most of the airfield decoys operated into 1941, by which time their need had greatly diminished. Once the bombing offensive against civilian targets intensified from late 1940, the need for a similar series of decoys for centres of population and production was realised. New decoys relying on the use of fires and simulated night-time suppressed lighting were constructed under the direction of Colonel Turner's Department at the War Office. Special Fires, SF, or 'Starfish' sites, under the control of No. 951 Barrage Balloon Squadron, were set up to reproduce fires, explosions, simulated street-lighting, and faulty blackout procedures. Most of Bristol's civil decoys were some way to the south and west of the city in order to attract incoming aircraft using the Severn Estuary as their guide. A series of six QL/QF sites at Burrington, in the Mendip Hills, extended the scope of Bristol's protection. It is, of course, as with any preventative measure, impossible to determine how successful these decoys were, but like Leighterton, and RAF Aston Down's 'Q' site at Horsley, Stockwood and Chew Magna both attracted large quantities of bombs. All these civil sites were operational into 1943 after which the main threat had receded.

Some decoys fulfilled slightly differing functions over time. Patchway began as a protection for Filton but subsequently had a wider remit as one of Bristol's ten civil decoys and then as a cover for railway marshalling yards. Important industrial sites and transport hubs were given specialised lighting effects: locomotive glows in railway sidings and docks; furnace glows in foundries; and appropriate lights in factories. Downside, Yeomouth, Severn Tunnel, and Patchway all developed from general civil decoys into sites with the railway lights and glows of marshalling yards (MY) and docks; Long Ashton and Stockwood became dummy factories (FL); and Severn Beach and Sheepway's decoy fires displayed the characteristics of blazing oil depots. Burrington's ambitious set-up aimed to represent Temple Meads, the Floating Harbour,

Canons Marsh, and assorted goods depots and sidings by replicating the layout on the ground and providing the appropriate lighting for the different elements. In Gloucester, Longney stood in for the Quedgeley depots with factory lights (FL); Standish mimicked the lights and glows of the marshalling yards (MY/LG); and Brimpsfield (MY/FL) shielded Cheltenham's numerous factories, depots, and railway yards.

In 1940, Hugh Casson took up the post of district superintending engineer for No. 9 Air Ministry Works Department, based in offices in Western Avenue, Cheltenham. His job, as an architect, was to over-fly airfields to assess their camouflage status. Along with colleagues, he developed techniques for obscuring and softening the hard shapes of airfield buildings, many of which had been designed applying the *moderne* crisp and sharp principles of Art Deco. A 1942 report on airfield camouflage described Filton and Hucclecote as 'conspicuous' with 'faded hedge painting'. Although the runways and aprons were 'not obvious', Filton was 'easily located by land-marks'. Hucclecote's factory buildings, it was suggested, might be improved by being painted as 'a housing scheme'. Whitchurch, on the other hand, was 'not very conspicuous' and its 'hedge painting was of good design and condition'. Staverton airfield was painted with lines to mimic hedgerows and irregular dark patches painted on the runways were meant to look like woodland with a fake rural lane snaking across the flying area. As well as airfields, Casson was involved in the defence of civilian targets which were vulnerable to bombing. In 1942, he produced a scheme for the city of Gloucester which superimposed organic and naturalistic forms on the regular street plan. This was apparently never implemented but he taught at Cheltenham School of Art and mounted an exhibition which demonstrated his camouflage protocols. The 'specially concealed normal factory' of Rotol at Staverton was cited, in

Birdlip W/T station.

correspondence with the Ministry of Home Security in 1943, as exemplifying the Directorate of Camouflage's repertoire. Sir Hugh would go on to teach at the Royal College of Art and design the Festival of Britain in 1951 and had been employed in an official capacity, but other architects, in the absence of private work, were forced to search out commissions. Erno Goldfinger's sole wartime contract was to camouflage the Imperial Tobacco cigarette factory in Bristol. The single-storey buildings at Smiths' Instruments new factory at Cleeve Grange were camouflaged using nets threaded with chicken feathers and with a dummy stream painted across them. If it was impossible to disguise some installations, it might be possible to hide them altogether. Diesel fuelled generators operated by the Pioneer Corps were used to produce smoke screens masking Avonmouth Docks.

The Germans used radio beacons transmitting from France to direct their aircraft onto targets by following beams. Masking beacons or 'Meacons', run by RAF No. 80 Wing popularly known as the 'Beam Benders', were set up to interfere with these transmissions and to give some protection to Bristol. The first Meacon in the area was installed at Templecombe (Somerset) with its associated receiving station at Kington Magna (Dorset). A further pairing was set up at Highbridge and Lympsham, both in Somerset, early in 1941. An emergency transmitter produced by the RN signals school was installed by RAF No. 80 Wing at the Birdlip W/T station.

The Royal Observer Corps (ROC)

At the beginning of the war, the county was covered by two Observer Corps Groups with a total of twenty posts, gaining the 'royal' honorific in 1941. No. 23 Group HQ in Bristol moved from Little King Street to Worcester Terrace in 1943, while No. 24 Group in Gloucester had its HQ in the GPO Exchange in George Street until 1942, when it was moved to Northgate Mansions. New posts had been added at Hill and Northleach early in the war and the Almondsbury and Dursley posts were both relocated during the course of the war. The official posts were supplemented by spotters occupying other vantage points such as hills, tall buildings, or church towers. Only minimal protection from the elements was supplied to some ROC posts in the shape of timber sheds to house the rudimentary equipment. Observers were expected to report incoming hostile aircraft to the central control which would then scramble fighter aircraft to intercept them. Schoolboys joined aircraft spotting clubs and became expert at aircraft recognition, often becoming unofficial observers until they were called up for military service. ROC duties went beyond passive reporting. From 1942, five posts, including Cleeve Hill, were equipped with Granite, which enabled them to project showers of red flares, to warn aircraft flying in poor visibility that they were close to high ground. As radar improved and it became increasingly possible to vector fighter aircraft onto incoming bombers, the ROC Groups were linked to Ground Control Interception (GCI) radar stations: Gloucester to Comberton (Worcestershire) and Bristol to Sopley (Hampshire).

The Balloon Barrage

Bristol was protected by barrage balloons organised as three RAuxAF squadrons, Nos 927, 928, and 929, each flying twenty-four balloons protecting Avonmouth and Bristol Docks. Although Filton airfield was not specifically covered, the presence of balloons in the vicinity would prove dangerous to pilots, particularly novices. A temporary HQ at 57 Victoria Street in Bristol was soon superseded by the Pucklechurch base camp which became No. 11 Balloon Centre. Here large purpose-built sheds were erected for storing and maintaining the balloons. From the end of August 1939, the three squadrons were deployed to their wartime stations: No. 927 Squadron to Avonmouth with its HQ at Twyford House, Shirehampton; No. 928 Squadron to Clifton with its HQ at 3 Caledonia Road; and No. 929 Squadron around the periphery of the city with its HQ at Victoria Street and a flight in reserve at Pucklechurch. Within a month, two of the squadrons had been sent to Queensferry on the Forth and to Felixstowe, leaving only No. 929 Squadron to man the port barrage. At the end of March 1940, a new No. 951 Squadron was formed at Pucklechurch to take control of the Bristol barrage, with its HQ in Caledonia Road. In May 1940, it was realised at Filton that the protection afforded by balloons far outweighed any benefit from flying fighters and a barrage was consequently installed by No. 935 (County of Glamorgan) Squadron. Flying would have to cease, and No. 2 EFTS moved to Staverton. At the same time, a barrage of twenty-four balloons was also located to the east of the Glosters works at Hucclecote, operated by 912 (County of Warwick) Balloon Squadron with its HQ at 'Woodstock', Barnwood. One of the sites, opposite Chosen Hill House, can still be traced by its earthwork perimeter bank, although the hut and other structures have gone. The barrage balloons were tethered to mobile winches by cables which allowed them to rise to 5,000 feet (1,550 metres). Their purpose was to force enemy aircraft higher, to make bomb aiming more difficult, and also to render them more vulnerable to AA fire. Many of the personnel involved were women of the WAAF, nearly 900 being allocated to Pucklechurch. Hydrogen was supplied by the ICI plant at Weston-super-Mare, providing the 19,000 cubic feet needed for each balloon. Only in early 1942 when the worst of the Bristol Blitz was deemed to be over was the balloon barrage reduced and the units reorganised. Permanent sites were shut down as a hazard to friendly aircraft and mobile units were formed, particularly to protect the cathedral city of Gloucester during the Baedecker raids. The two remaining balloon sheds and other buildings at Pucklechurch are rare survivals.

The Secret War

Communications

The dissemination of ULTRA intelligence produced by the decoders at Bletchley Park was the responsibility of MI6 which had a transmitting station at

Rendcombe airfield, North Cerney, near Cirencester, which had been set up by the wireless engineers of SIS Section VIII in September 1939. This Mark X transmitter posing as 'Radio des Beaux-Arts', ostensibly based in Brussels, was intended to communicate with stay-behind agents. This plan never came to fruition and North Cerney, whose transmissions could reach China and South America, was used to transmit ULTRA material from Windy Ridge at Whaddon (Buckinghamshire) which was Section VIII's HQ. Black propaganda was produced by the Political Warfare Executive based at Woburn Abbey (Bedfordshire), recorded onto 16-inch glass discs, and delivered to Cirencester by car, for transmission to occupied Europe. Some SIS Section VIII W/T operatives received their teleprinter training on a six-month course at Bristol Post Office, after which one of them was flown out to Lisbon on a flight from Whitchurch aerodrome. At Boddington, near Barrow, was an ATS-staffed army telephone exchange established in 1940 and guarded by troops of a Glosters' unit. It was later passed to the Royal Signals and then to the RAF. In 1940, the BBC was offered use of the tunnel which held the funicular railway linking the city of Bristol to its fashionable suburb of Clifton. Doubling as an air-raid shelter, the tunnel accommodated transmitters which were in daily use beaming programmes all over; and a studio, a recording room, and a control room which formed an emergency stand-by in case the local facilities were put out of action. Leckhampton Hill was part of the network of high-frequency direction finding (HF/DF or 'Huff Duff') stations. These structures would normally consist of an octagonal timber-clad tower set inside a brick blast-wall. However, an aerial photograph of the Leckhampton Hill tower appears to show an alternative construction, maybe brick or concrete, within the blast-wall. The tower stood within a square ditched enclosure with 200-foot-long (60-metre-long) sides with a tb crew-hut around 50 yards away outside the enclosure, all of which was demolished in the 1960s. Minchinhampton, which has also now vanished, has been suggested as another HF/DF station in the network. Chosen Hill, Churchdown, was the location for a BBC emergency transmitter which would only be brought into use after an invasion, accommodated in a small, stone building above the reservoir and manned by technicians who tested the equipment by briefly going on air each evening.

The Special Operations Executive (SOE)

Hatherop Castle at Fairford served as the Holding and Operational School (STS 45) for the Danish section of SOE and may also have involved Belgians. This was one of seven such establishments for the different national sections where agents went for the penultimate stage of their training. Individual agents might have received preliminary training at other SOE schools, and parachute training at RAF Ringway outside Manchester, and those who survived these experiences would then move on to one of these finishing schools. The Danish, Polish, Czech, and Norwegian Sections each had their own school where the basic SOE syllabus might be adapted for local

Hatherop Castle, a Special Training School for the SOE.

Forthampton House contains its medieval core and served as a Special Training School for the SOE.

conditions. On this stage of their training, agents were taught tradecraft, breaking and entering, and counterintelligence. During their stay, they would be despatched for further inputs of specialist training at other STS but would return to the holding school to consolidate their learning and put their new skills into practice. Hatherop Castle, based on the success of the commandant and the positive attitude of the Danish Country Section, was held to be the model upon which all the other such holding schools were based. Forthampton House, Tewkesbury, STS 49, formerly STS 45a, opened in January 1943, and is thought to have been used mainly by the Polish Country Section, as a complement to their HQ at Audley End in Essex. Here, agents could be totally immersed in an entirely Polish environment by way of food, language, dress, manners, and customs, in preparation for their perilous missions. The supremely hazardous nature of radio transmission meant that alternative forms of communication from agents in the field had to be employed. The Dickin Medal, the highest decoration in the animal world, was awarded to a pigeon bred in Cirencester and requisitioned by the RAF to be parachuted into occupied France several times, returning carrying messages from agents.

Depots and Supply

RAF Quedgeley, No. 7 MU, opened in April 1939 for storage and maintenance of aircraft equipment and motor transport (MT). It was on six discrete sites: HQ and Nos 1, 5, and 7 were on the former filling factory site (SO815135); No. 2 was at Hardwick (SO799121); No. 3 was in the north of Quedgeley (SO807148); No. 4 for aero engines, engine spares, and propellers was east of Hardwick (SO808124); No. 6 for MT was north of Haresfield church (SO808112); and No. 7 was newly constructed in December 1943, taking in stores from other existing sites as items were shifted around according to need. The HQ building is a standard 1938 design of a two-storey, neo-Georgian, H-shaped building with wings. The central block has six bays either side of a three-bay projecting entrance. From April 1944, Italian POWs from Yate were employed to labour on the sites and accommodated in tents at Quedgeley Court, and WAAFs from Innsworth were transported in daily. Shed No. 7 at Sharpness Docks was also hired; some stores, including power plants, were transferred to No. 11 Balloon Centre at Pucklechurch; and throughout the war the original buildings were supplemented by Blister and Bellman hangars. The army supply or mobilisation depot at Northfield, Tetbury, had been set up pre-war to store AA guns and equipment for the TA units which would provide the nation's protection against aerial bombardment. It continued in use as a RASC Depot. The US Army established an engineer depot (G-35) in part of the Elizabeth Shaw chocolate factory in Greenbank, Bristol.

To a greater extent even than in the First World War there was an insatiable need for the storage of munitions, particularly bombs and ammunition. In the hunt for suitable facilities for the army's reserves, Hasley and Avening quarries near Nailsworth were considered but rejected in favour of the extensive complex at Corsham in Wiltshire. The RAF managed to acquire three disused railway tunnels, aligned to the long closed GWR line from Monmouth to Coleford, lying between Redbrook and Newland in the Forest of Dean. These air ammunition parks were intended to take the pressure off the central depots which found it difficult to receive and issue at the same time. They would serve the new fighter stations in Wales and the south-west established as a result of the *Luftwaffe*'s presence in north-west France. RAF Newland opened as No. 56 MU but was subsumed into No. 11 MU at Chilmark the following year. The depot's administration was housed in Newland Station and staff were billeted in empty industrial premises. The easily accessed Newland Tunnel was suitable for munitions storage but the other two, facing each other, were less so. Narrow-gauge lines were laid, and a road transport loading dock was contrived. The narrow confines of the tunnels precluded the storage of large bombs and the site would never really function effectively. In June 1941, HQ RAF No. 42 Group had considered Clearwell Cave as a possible storage site for obsolete bombs which were the cause of particular problems of decay and instability. Next, the possibilities of using Blakeney Walk, Lower Soudley, and Russell's Enclosure were explored, only to find that the War Office had got there first, using the last named for the storage of chemical weapons. The War Office had also secured the use of the Moseley tunnel for ammunition storage which, along with the Hawthorn tunnel between Drybrook and Cinderford, stored naval munitions. The RAF would return to the Forest of Dean in 1944, securing roadside dumps at Blakeney Walk and Lower Soudeley which, together with the tunnels, stored 2,000 tons of bombs. In 1945, with the removal of the holdings to RAF Rhoose in Cardiff, Newland could finally be closed. Cinderford became one of the three largest ammunition storage depots in England for the US Army with the main site at Acorn Patch to the south of the Speech House where chemical weapons were stored. Other US sites were Worral Hill, Worcester Walk, Lydbrook, and Saintlow Enclosure. Ordnance was stored in large containers or under sheets of corrugated iron, either alongside roads, or deep into forested areas using the forest's natural cover. One of the units tending these dumps was the US Army 596 Ordnance Ammunition Company which arrived in England in September 1943 and was billeted in camps at Lower Lydbrook and Broadwell until July 1944 prior to being deployed to France. These US technicians and those such as the 78th Ordnance Depot Company at Ashchurch worked alongside artificers, packers, and craters of the RAOC in order to prepare munitions for issue to the forces sailing for France on D-Day.

The US Army SOS, controlling the supply element of US forces in the British Isles, under General J. C. H. Lee, established its HQ in Cheltenham with stores depots at Gloucester, Bristol and Lydney Harbour. General Lee's HQ occupied

Ashchurch Depot used by the RASC and the US Army Service of Supply.

Lydney Harbour, the HQ building of the US Army Service of Supply depot.

temporary office buildings (TOB) composed of interlinked prefabricated hutting, previously commissioned by the authorities and earmarked for the possible evacuation of the War Office. Benhall Farm became HQ US Army Quartermaster Corps and use was made of Staverton Airfield, Cheltenham Boys' GS, and Cheltenham Ladies' College. A US Ordnance SOS Company which arrived in Britain in July 1942 took over part of the recently built vehicle supply and maintenance depot of the RASC at Ashchurch, near Tewkesbury, and located there in order to be in a convenient location for shipping in and out of Bristol. It was situated alongside the railway connecting to Birmingham, Gloucester, and Evesham, with extensive sidings serving the depot. Aerial photographs, taken by the RAF in 1947, show eight sawtooth-roof sheds and three flat-roof hangars of similar size, with numerous smaller hangars, together providing 1.75 million square feet of space, a generous nearly 20 per cent of which was allocated to the US unit. A large accommodation camp sits alongside with an adjacent POW camp. This depot became a key component of Operation Tilefer, the supply of boxed and cased vehicles to the British Isles. The largest of the US Army's munitions depots was relatively close at Kingham on the Gloucestershire–Oxfordshire border. Lydney Harbour US Army SOS Depot opened in October 1942, and by the next year it was using 60 per cent more covered accommodation than it had initially been assigned and almost all of the open storage area, together covering 550,000 square yards. A year later, it was operating at 100 per cent but on a reduced area of 100,000 square feet. After D-Day, the site became a Central Salvage Repair Depot. Now an industrial estate, the Guardhouse, HQ building, a row of Nissen huts and several large warehouses remain. The area was home to other maintenance and supply units including RAF No. 238 MU, the Marine Craft Salvage Unit at Tewkesbury.

Hospitals

Typical of the many US hospitals in the UK, especially in the Gloucestershire area, was No. 188 General Hospital at Cirencester. On arrival on the RMS *Mauretania* in April 1944, the unit initially established a HQ in a 'four-storey building on the corner of Southgate Street and Longsmith Street', Gloucester, with an officers' mess in Wesley Hall, Sidney Street, and mess halls for the enlisted men in Reservoir Road and Sisson Road. The hospital opened in Cirencester Park in May 1944 with 1,082 beds in Nissen huts and the capacity to expand with a further 435 beds in tented accommodation. From 14 June, casualties began arriving from Normandy along with several hundred wounded POWs. Despite latterly specialising in neurosurgical cases, the hospital treated almost 11,000 patients with around 90 per cent being successfully discharged, many of them returning to duty. From March 1944, the hospital had been one of fourteen

Northwick Park, three of the thirty odd huts which, along with numerous Nissen huts, made up the former military hospital.

such units administered by the 15th US Hospital Centre (later the IVth Hospital Group) which included the 110th at Ullenwood and the 111th at Daglingworth.

Frenchay Park, converted as an emergency hospital for victims of the Bristol Blitz in early 1942, became the nucleus of the US 298th and, later, the 100th General Hospital. The US Army 6th Field Hospital was in Avonmouth, and the 600-bed Emergency Medical Services (EMS) Centre at Bristol was allocated to the US Army in May 1942 as the US Western Base Section General Hospital, with the 834-bed EMS at Cheltenham as the US Southern Base Section General Hospital. In the build-up to D-Day, over fifty US military hospitals were earmarked for the receipt of casualties from Normandy. General Hospitals included the 40th at Daglingworth; the 62nd at Ullenwood; the 94th at Tortworth Court; the 160th at Stowell Park, Northleach, where ground-breaking work in heart surgery was pioneered; and the 188th and 192nd at Cirencester. The 327th General Hospital was at Northwick Park, and five hospital trains were held ready in Batsford Park with four more in Cirencester. In Fairford Park, No. 32 US Station Hospital was built, later becoming, as did several other such hospitals, a Polish Resettlement Corps camp whose huts were finally demolished in 1977. The majority of these US military hospitals were accommodated in collections of Nissen or Quonset huts connected by covered walkways, with distinctive water towers and brick generating stations.

British military medical provision included the Army Blood Transfusion Service in Bristol and the HQ of Queen Alexandra's Imperial Military Nursing Service

(QAIMNS) in Cheltenham, 1942–45. Down Ampney airfield was designated post D-Day as a central air evacuation base, and RAF Innsworth had a hutted RAF Station Hospital, 1940-48. As in the Great War, many country houses were made available as convalescent homes or auxiliary hospitals, often run by the British Red Cross. These included Little Sodbury Manor; Brockhampton Park at Andoversford; Cirencester's South Cerney House; Edgeworth Manor, Stroud; Hartpury; Estcourt at Tetbury; and Dockem House, Coates.

Industry

Beyond the aircraft factories, which have been dealt with above, there were many further industrial enterprises which all contributed to the war effort. Workers and their families coming into the county with their employers' shadow factories made a social impact, at Stonehouse, the children of war-workers doubling the school roll. In the Forest of Dean, auxiliary pioneers were recruited from foreigners originating in belligerent nations, dubbed 'The King's Most Loyal Enemy Aliens'. One company, the 220th, worked to establish sawmills, expanding the output of home-grown forestry.

Munitions

Many factories whose peacetime markets had disappeared found themselves suddenly part of the munitions industry. As rationing reduced the production of chocolate, Weber's factory in Fishponds, Bristol, was taken over by Thrissells Engineering for the manufacture of breech casings and barrels for 20-mm Oerlikon LAA cannon. Other firms simply switched their entire production to munitions. Newman's engineering works in Yate produced shell casings. Berkeley Primary (formerly Secondary) School Science laboratories, wood, and metal workshops were taken over by Listers of Dursley making diesel engines which were stored in the Dairy Institute alongside a sheet metal company evacuated from Eastbourne. As well as their vital ball-bearings, Hoffman's also manufactured 0.303-inch bullet cores for onward transmission to ammunition factories. While rifles were manufactured and assembled at ROFs, many of the smaller parts were produced by hundreds of smaller companies. In 1936, Bristol Engineering Manufacturers (BEMA) had been formed with a dozen firms signing up to co-operate in obtaining commissions and during the war BEMA acted as a clearing house for contracts. One of its members ESS (Signs) Ltd manufactured magazines, trigger guards and swivels for Lee Enfield 0.303-inch rifles. Smiths' war production of instruments also extended to mechanical time fuses for anti-aircraft shells. Fielding and Platt in Southgate Street, Gloucester, made presses for use in the manufacture of shell casings. The Gloucester Railway Carriage and Wagon Company built specialist rolling stock for transporting tanks; over 750 Churchill tanks; Bailey bridges; and parts for the Mulberry Harbours.

Railways

The Great Western Railway (GWR) with its regional office at Northgate Mansions, London Road, Gloucester, had identified those installations in the network most vulnerable to bombing or sabotage. These included the Severn Tunnel with its associated pumping stations and Chepstow Bridge. Temple Meads station in Bristol was hit particularly badly in January 1941 but even light damage, if caused in the wrong place, could paralyse the rail service and a single bomb, which took out a railway bridge in Cheltenham in December 1940, caused a blockage which took a week to clear. Also recognised was the importance of rail links to sites of military importance. The East Gloucestershire Railway Company had been formed in 1861 to link Witney and Cheltenham via Fairford. In the event, only the 14-mile (22.5-km) stretch from Witney to Fairford was completed and absorbed into the GWR in 1890. This branch-line became particularly valuable during the war as it carried munitions and personnel to the airfields at Southrop and Fairford itself. The line from Yate to Thornbury through Iron Acton and Tytherington was closed to civilian traffic for the duration, and most of the lines in the Forest of Dean and those round Gloucester Docks were confined to freight traffic. Stations at Andoversford and Dowdeswell, serving the RE Depot, were closed to passengers, and a new halt at Chittering, north of Avonmouth Docks, was opened for the sole use of munitions workers. According to a 1942 'Secret' list of government wartime railway arrangements, dedicated sidings were provided for the RAF airfields at Aston Down, Babdown Farm, Little Rissington, South Cerney, Innsworth/Staverton, Whitchurch, the Pucklechurch Balloon Depot, and the Quedgeley Depots. Sidings were provided for the army at Little Badminton Camp; Blaise, Cribbs, Markham, and Purdown AA Camps; Benhall Farm, Cheltenham; the Beachley Technical School; Chipping Campden militia camp; Reservoir Road Camp; the Monk Meadow ordnance depot at Gloucester Docks; and the Tetbury RASC Depot. Government supply facilities such as the Avonmouth Cold Store, the Badminton Store, the Ministry of Supply salvage depot at Lydney, and the Tewkesbury food store were all given sidings. While the Yate aircraft factory was conveniently situated near the railway, for other industrial premises such as the Gloster's aircraft works at Brockworth and the MAP Aeronautical Panel Plywood Co. at Lydney, new sidings would be necessary. From 1940, with War Office approval, the railway companies set up an emergency radio network for use if the telephone and telegraph systems were to fail. Radios with a range of 50 miles (80 km) were installed in sixteen fixed stations across Britain, one of which was Gloucester, along with one of the additional dozen mobile stations mounted on 30-cwt Karrier Bantam trucks. A rail mobile unit was stationed at Severn Beach. Gloucester was also provided with a Type B2 protected control room with reinforced concrete walls and a roof 15 inches (38 cm) thick. This had seven desks, a battery room, generator room, and a telephone and switch room. It was entered through a door at each end with

Above: Hallen, the HQ building of the Air Ministry fuel distribution system in the Avonmouth area.

Right: Tewkesbury, one of the characteristic 'stiles' which marked the course of fuel pipelines.

an airlock and steel blast doors. As part of ARP, both Bristol and Gloucester were provided with back-up telephone and control desk facilities, in order to keep the vital railways running.

Fuel Supply

Most aviation fuel arrived, ready refined, by ocean tanker. From the Avonmouth storage depot, it would travel by rail-tankers to the AFRDs; these would, in turn, use the railways to supply the AFDDs which would then generally use road tankers to deliver fuel to each of the ten or so individual airfields they served. This system, involving continual loading and unloading, tied up large numbers of rail and road tankers and, owing to the enormous quantities of fuel needed for ambitious and frequent air operations, became quite difficult to sustain. In 1941, a 100-mile-long (160-km-long) pipeline was constructed linking Avonmouth and Walton-on-Thames. Eventually, by 1942, a second pipeline joined Avonmouth to Stanlow on the Mersey and, by 1944 this had been extended eastwards from Stanlow, to create a ring-main which ran over the Pennines to Misterton in Nottinghamshire, and then south, via Sandy in Bedfordshire, to Thames Haven. A spur via Saffron Walden served the bomber fields in East Anglia and another ran south from Sandy to Aldermaston and onwards to Fawley and the Isle of Wight to feed PLUTO (the Pipeline Under the Ocean) on its way across the Channel to France. In the build-up to D-Day, before the fuel pipelines had been completed, hundreds of trains carried both bulk fuel in tanker trucks and filled jerricans in goods wagons from the Avonmouth depot to the invasion ports. The aerodrome at Whitchurch was supplied with fuel direct from Avonmouth.

From the 1920s, a fleet of tanker barges had begun delivering oil from Avonmouth to Gloucester with storage tanks at Monk Meadow Dock and a War Department fuel depot at Hempstead Wharf. The former Quedgeley Fuel Depot on Wharfdale Way on the Sharpness Canal probably represents the Gloucester AFRD which served local airfields around Gloucester and Cheltenham. The AFRD at Henbury or Hallen, on Hallen Road, was served in 1939 by a railway siding on the line that runs directly from the oil storage tanks which line the riverbank in Chittering and later by a spur from the Avon-Thames pipeline. It is still in operation and the Avonmouth area control building still stands alongside the discernible outlines of ten 2,500-ton storage tanks. About a mile to the north on Berwick Lane is the Berwick Wood storage depot which marks the junction of the Avon-Thames (A/T) and the North-South (N/S) pipeline from Stanlow, forming one corner of the ring-main. Hallen served the airfields at Filton, South Cerney and Aston Down, while Little Rissington, Fairford, and Southrop got their fuel from the Islip AFDD, over the border in Oxfordshire. The pipelines were marked above ground by white-painted 'stiles'.

POW Camps

POW camps were established all over the county, increasing in number as the war progressed, and often in repurposed premises. Springhill Lodges near Chipping Campden began life in 1939 as a militia camp. It was occupied by US forces who vacated it in 1944, and then became POW Camp No. 185. Similarly, Northway Camp at Ashchurch was a US Army camp from 1942, being given over to POWs in 1945. It consisted of nineteen timber huts accommodating 800 German POWs, captured in France after the Normandy landings. A wire fence with guard towers was built around it, and this secure compound was linked by a footbridge over the railway to the Ashchurch Depot where the prisoners were put to work. Bourton-on-the-Hill Camp comprised a total of thirty-five barrack huts in six discrete compounds, each surrounded by double-wire perimeter fences and watchtowers. There was a theatre *cum* chapel with simulated stained-glass windows. Northwick Park Camp began life as a US Army hospital prior to its conversion into a POW hospital camp, late in 1944. The camp at Churcham started as a searchlight site and shares the same style of buildings, BCF huts, as another nearby camp at Naas House, Lydney. The Naas site had been requisitioned by the Ministry of Supply in 1939 and a total of thirty-five huts, a mix of BCF and Nissen, were erected along with tents for troops. It then served as a camp for German and Italian POWs, working at the SOS Depot and on farms. The camp opposite Woodfield Farm in Churchdown, was first occupied by the Pioneer Corps servicing smoke generators around Cheltenham and Gloucester before becoming a POW Camp in 1943. It consisted of small groups of huts dispersed around a field. In contrast to these stereotypical POW Camps, there were also less stringent regimes. The camp on Gloucester Road, Thornbury, was for just thirty-forty Italian POWs working on local farms, largely unsupervised. Many such labourers were quietly integrated into the community and in the Forest of Dean, probably Coleford, the Italian POWs fielded a football team against local teams. Leckhampton Court Camp, another former US Army camp, became a German POW work camp and may have included tented accommodation beside the huts. As in the Great War, there was a camp at Sedbury, on the Beachley peninsula, whose remains may possibly still be visible in Grahamstown Road.

Civil Defence

Cirencester was selected to run practice 'Black-outs' in July 1939, with the whole of the county following suit in August and the ARP organisation taking shape over the next few months. Gloucester's ARP report centre was in the basement of the Guildhall and the county control centre was at Shire Hall under the direction of Major Sir Frederick Cripps, the county controller. Gloucester Rugby Club's

Kingsholm ground was requisitioned for use by the Civil Defence organisations, being used as an equipment depot and for training and public demonstrations of equipment and procedures. Cheltenham's ARP control centre was in the basement of Shirer & Lance's department store in The Promenade with an air-raid siren mounted on the roof. Bristol's ARP control was set up in 1938 at 55 Broadmead with a reserve control centre in the university. The area was split into six operational divisions each with its own organisation but co-ordinated from central control. In September 1941, the central control moved to Badminton Junior School on Cote Road in Westbury-on-Trym, and Broadmead became the reserve, with a second reserve in the university's tower basement. Bristol had nearly 200 manned wardens' posts and a special observation post was set up in the tower of the Royal Fort. The city's Fire HQ was in Rupert Street. The cellars under 19–21 Woodland Road, Tyndall's Park in Bristol, housed the HQ of General Sir Hugh Elles, the commissioner for No. 7 (South-Western) Region. Here, the war room was strengthened with steel girders and equipped with steel doors, priority telephone lines, a gas filter, and escape hatches. The ambulance control centre for south Gloucestershire was at the old workhouse in Chipping Sodbury.

Air-Raid Shelters (ARS)

Government guidelines dictated that public air-raid shelters (ARS) should be provided for 10 per cent of the population who might be caught away from home or work during an air raid. Places of work must therefore provide sufficient shelter accommodation for their workers, and householders should ensure that shelter was available for their families. Trench shelters for 40,000 people were provided in Bristol and these were supplemented by communal brick surface shelters, cellars and basements. The disused Port & Pier Tunnel, constructed in the 1860s for an uncompleted railway running under Clifton Down connecting Bristol to Avonmouth, was used as an ARS. There were 234 recorded ARS in Gloucester including many communal fifty-person ones, two of which survive in Brockworth used as garden sheds. A shelter at Cirencester Brewery was constructed of Norcon concrete tubes 6 by 4 feet (1.8 by 1.2 metres) and 24 feet (7.4 metres) long, seating thirty-five people, under 3 feet (0.9 metres) of soil and equipped with electric lights. In Bristol, well over 5,000 men and women were trained as air raid wardens and were based in 187 posts, many of them in church halls or schools but later on in purpose-built brick and concrete cabins. Some organisations would take the coming emergency more seriously than others. On 1 September 1939, the minutes of the Emergency Sub-Committee of the Midcounties Co-operative Society recorded a resolution to turn the basements of four of the society's stores in Gloucester and Cheltenham into air-raid shelters, and that the society's architect be briefed to that effect. Two meetings later saw a decision to provide staff with training in first aid and firefighting. Offices would

Falfield Park, Eastwood, begun *c.* 1820 for the earl of Liverpool, one of the two national ARP/CD schools.

be set up to deal specifically with rationing measures, initially with regard to fuel. At least four schools in Bristol were provided with cut-and-cover ARS, recently (2020) uncovered and filled in with concrete as being dangerous.

Fire and Rescue

The fire service in Bristol began with three fire stations: Bridwell (HQ), Avonmouth, and St George plus two fire floats in the port. By the end of 1940, the AFS had twenty-six fire stations, more fire floats, and mobile units. The NFS, which was created in August 1941 to facilitate joint planning, standardised equipment, and integrated operations, established Bristol as the centre of No. 17 Fire Force with its HQ at 'Crete Hill', Westbury. Static water tanks were established, supported by 25 miles (40 km) of 6-inch (15-cm) steel pipelines around the city to supply water even when the mains had been fractured. Gloucestershire Constabulary recruited 800–1,000 special constables who were on duty every night between 6 p.m. and midnight. In the early days, they guarded VPs, but by early 1940, they were settled in as auxiliary police with support from a Women's Auxiliary Police Corps organised into clerical and transport sections. Each town established its own depot, Cirencester's being at Querns Hill. Eastwood Park at Falfield had been set up by the Home Office in 1936 as a national ARP facility, primarily for anti-gas training. The Sir Thomas Rich School in Gloucester was also used by ARP for anti-Gas training courses. The REs' No. 7 Bomb Disposal (BD) Company was based in Bristol and ran a school which trained the US Army's

234th BD Company. The experience of Bristol in particular was to demonstrate the importance of all this infrastructure.

Air Raids in Gloucestershire and Bristol

Bristol suffered ten major air raids with over 1,200 tons of HE or oil bombs and 250 tons of incendiaries being dropped. Nearly 1,500 people were killed, and 90,000 properties were damaged or destroyed. A mortuary was set up in St Aldhelm's School Rooms in Bedminster, with an emergency reserve at St Agnes Church Hall in St Paul's. Gloucester and Cheltenham were both bombed between December 1940 and April 1942 causing several hundred deaths, and in Cheltenham, five children were killed playing with an unexploded bomb. We have seen how the aircraft factories at Filton, Brockworth, and Yate were bombed with consequent fatalities and injuries, but there were other instances of the military and the ARP services suffering casualties. A raid on 'T' Camp, St Andrew's Road, Avonmouth, on 15–16 September 1940 killed three soldiers serving with the Glosters' 8th (Home Defence) Battalion. A fortnight earlier, a Home Guardsman had been killed in a raid in Portishead, and a week later fifteen members of the 13th (City of Bristol) Battalion would die in a raid on the Filton area. In that same raid, thirteen soldiers of 4th Battalion, Berkshire Regiment, from 'T' Camp, on duty at the Filton Aeroplane Works, were killed along with a member of a searchlight battery. Servicemen returning from or going on leave were killed, particularly on the night in December 1940 when Temple Meads station was bombed, taking the lives of nine soldiers and three airmen. On another night that same month, the deaths of eleven ARP wardens and two of their messengers aged fourteen and sixteen were reported. On yet another night, five firemen died. Men and women ARP wardens, barrage balloon and bomb-disposal personnel, police officers, Home Guardsmen, including at least one sixteen-year-old, women casualty service workers, WVS members, seamen on ships in Avonmouth Docks, and fire watchers were all among the fatalities caused by the air raids, but the vast majority were families of civilians, bombed in their homes. Thornbury was bombed on 116 separate occasions, receiving eighty HE bombs and large numbers of incendiaries.

Evacuation

Cirencester Corn Hall was the reception centre for evacuees from London with around 500 arriving in Cirencester and over 1,000 in Gloucester, contributing to the total of 4,500 across the county. Many of these drifted home during the Phoney War, but by the summer of 1940, the total had climbed back up to 5,000. Whole schools were also evacuated. Dursley GS found itself sharing with the Moseley School of Art from Birmingham and Harrogate High School, all three schools alternating half-day use of the classrooms. Following the unsettling experience of being bombed, interpreted as being intentional, Clifton College was evacuated to Bude in Cornwall.

Brislington, the gatehouse building of the government's Temporary Office Buildings estate for civil service departments evacuated from London, similar to other such estates in Gloucester and Cheltenham.

Prior to the outbreak of war, government departments had planned to decentralise much of their work to supposedly safer areas outside London. While the royal family would go to Worcestershire and the Admiralty to Bath, as part of the wider operation codenamed 'Black Move', Gloucestershire and Bristol figured strongly in these plans, with the Cabinet bound for Cheltenham. A significant number of the Air Ministry's civil servants was destined for the county, with very precise numbers of personnel to be distributed: 209 to Bristol (AJ); 602 to Cirencester (JK); 1,055 to Gloucester (CJ); 734 to Stroud (QA); and 634 to Tetbury (JJ). Not all of these plans were implemented, but the Air Ministry civil aviation staff made it to Bristol, and Stroud played host to the accounts branch occupying Lansdowne Road School of Arts & Crafts, the Imperial Hotel, the Holloway Institute, the George Street Subscription Rooms, and the Bear Inn at Rodborough. Additionally, parts of the Ministry of Home Security went to Gloucester, while the War Office established an annexe code-named 'Smith' in 1940 at Chavenage Manor near Tetbury and claimed accommodation in Cheltenham. As well as taking over existing buildings, the authorities commissioned TOBs, six being constructed in Gloucester for AM staff and a further twelve in Cheltenham for the War Office. Those at Benhall Farm would later be occupied by the US SOS HQ. Bristol supplied two TOBs at Concorde Drive, Westbury-on-Trym; another on Gill Avenue, Fishponds; and two more at Flowers Hill, Brislington.

As well as civil servants, some of the nation's other treasures were evacuated. Parts of the Tate Gallery's art collection were taken to Sudeley Castle for safekeeping before being moved on to the more secure Manod slate quarry in Ffestiniog in North Wales. On the suggestion of the University of Bristol who had been seeking a safe store for its archives, Westwood Quarry near Bradford on Avon also stored artefacts from the British and the Victoria & Albert Museums. Local masterpieces were also safeguarded, and the stained-glass windows of Fairford Church were removed and stored in the cellars of Fairford Park.

Uniformed Organisations

For many women who joined the WLA, their introduction was a two-week course at the Royal Agricultural College, Cirencester. For others, it was a 6 a.m. start on a freezing cold morning getting the cows in for milking. By the end of the first year, there were 300 Land Girls working in Gloucestershire, a number which would increase throughout the war. Training took place at different locations including Cirencester Park, while the WLA County Office was at Winterbourne Park. Accommodation could be in a country house like Dumbleton Hall, Southam Priory, or a wing of Miserden Park House, or it could be in a purpose-built hostel as at Berkeley, Tetbury or Oaklands Park in the Forest of Dean, each of which housed forty girls in a standard hutted layout. Two-thirds of the hostels were run by the WLA organisation itself with paid resident wardens, but one in five was run by the YWCA. Land Girls worked on all sorts of farms and some, the lumber jills of the Women's Timber Corps, worked in forestry.

Many women joined the Women's Voluntary Service (WVS) to run canteens, welfare services, or to organise evacuation operations. In the Cirencester area, the WVS HQ was Ampney Park, home of Lady Cripps. Recruitment to youth organisations had surged in the immediate prelude to war with many youngsters joining the Red Cross and St John's Ambulance to help ARP activities. Those boys at Cheltenham GS who belonged to the local Scout Troop were volunteered as ARP messengers, but by 1941, they would be enrolled in ACF, ATC, or Sea Cadet Corps contingents in preparation for conscription. At Farmors School, Fairford, an ATC flight formed in May 1941, and by the next year, the Girls Training Corps branch had formed there with forty members.

8

The Cold War and Beyond in Gloucestershire and Bristol 1945–2020

After the war, BAC was supplied with aluminium by the Ministry of Supply in order to build prefabricated houses and schools from aluminium panels, helping to solve the chronic housing shortage caused by the bombing. Uni-Seco, a company that had produced hutting for use on airfields, supplied single-storey prefabricated housing to Bristol. Alongside BAC, the Hawksley company's shadow factory at Hucclecote also manufactured a range of aluminium single-storey housing with many, despite only being expected to last around ten years, still standing in Podsmead. They also built some experimental two-storey houses. The British Iron & Steel Federation (BISF) erected a small number of two-storey houses, seen as being permanent dwellings, some of which remain in occupation.

In view of the apprehensions raised by the onset of the Cold War in the late 1940s, Eastwood Park Training School, Falfield, continued in service under the auspices of the national Civil Defence organisation. It would boast a simulated village, 'The Range', for search and rescue training; for experiments in protection against radiation and investigations into the properties of different materials subjected to blast or radioactive exposure; and for developing fire and rescue techniques. By the late 1960s, there was a general consensus that the very notion of the population surviving a nuclear attack was untenable, and in 1968, it was closed down. Until this time, incidentally, the Polish Resettlement Camp in the former Daglingworth Barracks had boasted the only Polish CD unit in the country, led by a former paratrooper, representing Gloucestershire in competitions, and consistently winning trophies through 1959 and 1960.

Throughout the Cold War, there were constant changes of policy, partly as a result of shifts in international relations; partly due to technological advances; and partly down to differences of opinion between political parties. The protected structure at Ullenwood (S0936174) neatly illustrates some of these changes. It began life on the

Ullenwood, the bunker which served as an anti-aircraft operations room, controlling the Gloucester/Brockworth GDA until 1954; then the Civil Defence Southern Region Group HQ until 1962; then a sub-regional control for the Home Office and a Regional Government HQ until 1985.

site of an army camp and a US military hospital as an anti-aircraft operations room (AAOR) in the rotor system, operating as such to control the Gloucester/Brockworth GDA until 1954. Next, it became a Civil Defence Southern Region Group HQ. After that, in 1962, conversion as a sub-regional control alongside Bolt Head in Devon, brought it into the sphere of the Home Office's plans for post-Armageddon seats of government. Following a spell as a regional government HQ, it was superseded in 1985 by an alternative purpose-built installation. The drama of its official life was finally concluded after time as Gloucestershire County Council's Main Emergency Planning HQ. It then assumed a more prosaic role as a storage facility for counterfeit goods seized by the county's Trading Standards department and now (2021) has been spruced up and graces the grounds of a new country house.

The Army in Gloucestershire and Bristol

Local Units

The end of the war saw a general reduction in the armed forces, although conscription would remain into the 1950s in the form of National Service. The Gloucestershire Regiment was reduced to one Regular and one Territorial battalion. The 1st Battalion fought in the Korean War and was part of the brigade group that held off overwhelming numbers of Chinese troops at the Battle of the Imjin River in 1951, winning two VCs in the process. Reduced to just sixty-three effectives and soon to be dubbed the 'Glorious Glosters', the battalion had lost over fifty killed and more than 500 taken prisoner. The regiment served all around the Commonwealth and completed seven tours in Northern Ireland between 1968 and 1991. The TA was reconstituted in 1947 to include the former 5th Battalion, but it was absorbed into the Wessex Volunteers in 1967. In 1994, as part of the peace dividend arising from the break-up of the Soviet Union, the Glosters, after 300 years' service, were amalgamated with the Duke of Edinburgh's Royal Regiment to become the Royal Gloucestershire, Berkshire and Wiltshire Regiment. This was then, in 2007, merged into The Rifles, whose 1st Battalion's depot is currently (2021) Beachley Camp, which is scheduled to close in 2027. The RGH Yeomanry became an armoured car regiment until, reduced to a cadre in 1969, it provided two squadrons of the Royal Wessex Yeomanry, formed in 1971, by combining the Yeomanry regiments of Devon, Dorset, Gloucestershire, and Wiltshire. Their 'C' (RGH) Squadron currently (2021) provides training for reserve crews for Challenger 2 battle tanks, maintaining a presence in the county at the Cirencester ARC. The former 76th and 98th HAA Regiments (TA) went through a number of reorganisations and amalgamations with other local AA units and have effectively been bred out of the system, but a vestigial successor unit, 266 (GVA) Battery, 104 Regiment RA, equipped with 105-mm light guns and operating unmanned aircraft, is based at Whiteladies Artillery Grounds.

Barracks and Military Camps

Reservoir Camp had been created on the north-east side of Robinswood Hill at Matson outside Gloucester as a hutted camp in 1939 to serve as a depot for the Gloucestershire Regiment. Although National Servicemen were still being taken in, by 1947 the camp authorities felt able to sell off a timber hut to the Catholic Church for resiting in Southfield Road. This hut, which cost them just £15, would serve its community well until 1966. The new Robinswood Barracks was built on the site of the camp in 1954–55 and its last occupants were redeployed to Worcester in 1968 after which the camp was largely covered by new housing, while part of the site became a Country Park. Horfield Barracks, decommissioned after the war, was largely demolished in 1966. The site became a GPO telephone engineering works and was then developed for housing in 2000. Parts of the outer wall and the chapel survive. The Army Technical School (Boys) at Beachley became the Army Apprentices School in 1947 and was upgraded as a college in 1966. It was dedicated to producing technically qualified soldiers as potential officers or NCOs in the technical corps, chiefly RE, REME, and RAOC, with graduates of this apprenticeship scheme going on to achieve the rank of major-general. The college closed in 1994 when the scheme was discontinued.

Along with 1st Battalion, The Rifles at Beachley, other regular Army deployments in the county include the HQ 104 Logistic Brigade at South Cerney, and HQ 1st (UK) Signal Brigade at Imjin Barracks, Staverton, the former RAF Innsworth, also accommodating a Ghurka battalion of NATO's Allied Rapid Reaction Corps. Along with the Yeomanry at Highfield House in Cirencester,

Army Reserve Centres in the county include: Whiteladies Road Artillery Grounds in Bristol, formerly home to 100 Regiment RA which has (2021) been placed in suspended animation, being replaced by 266 (GVA) Battery, alongside the Bristol University OTC; Horfield Common in Bristol, which suffered a lightning strike in 1953, and now home to Military Intelligence and Signals units; Arle Road in Cheltenham hosting a branch of the Joint Cyber Reserve Force, created in 2013; and Gloucester, a signals unit. A reserve company of The Rifles is based across Bristol, Gloucester, and Hereford, with a detachment of the Wessex Field Hospital in Gloucester. Elements of REME theatre support companies are based in Bristol and Gloucester. Activities at the post-war drill hall at Cinderford are now confined to the ACF. Over twenty contingents contribute to the three companies of the City and County of Bristol ACF and a further eighteen belong to the Gloucestershire Cadet Battalion (The Rifles), scattered widely around the county. Pates GS in Cheltenham, with 200 cadets, is one of those schools which maintains a strong CCF contingent, originally formed by the merging of their ACF and ATC elements in 1955. Other CCF contingents can be found in schools in Cheltenham, Gloucester, Lydney, Stonehouse, Tewkesbury, Coleford, and Bristol.

Cinderford, the 1957 drill hall of the Herefordshire Regiment now used by the Army Cadet Force.

Air Defence

After the war, Britain's existing AA capability was reorganised as Rotor, a network of radar stations, ROC aircraft spotting posts, GDAs equipped with batteries of radar-controlled HAA guns and fast interceptor jet fighters.

Anti-Aircraft Defences

The rotor system was designed to meet an ever-evolving aerial threat, generally seen as emanating from Russia. A Sector Operations Centre (SOC) was the hub for the operation, that serving the south-west, No. 12 Group, being located in a former quarry, previously used for ammunition storage in the Second World War, at Box, near Corsham in Wiltshire. At the beginning of 1946, the RAF's only two operational squadrons of Gloster Meteors were stationed at Colerne, also in Wiltshire, but were soon redeployed to East Anglia. No. 501 Squadron, flying Vampires, was based at Filton until the RAuxAF was disbanded in 1957. The GCI radar stations, in the case of this region at Comberton (Worcestershire) and Charmy Down (Somerset), would provide the necessary data for the SOC to direct the AA guns and the fighters. Given the prospect of enemy bombers flying at 80,000 feet (24,600 metres) at a speed of 700 mph, it was clear that the existing wartime systems needed to be updated, but new equipment and communications systems would be slow to come on stream. Although the same guns that had finished the war were still in use, they had been much improved with automated loading and were laid electrically to radar generated data. They continued to be organised in GDAs controlled from AAORs. Group 2 which was aligned with the No. 12 Group SOC included the Bristol and Gloucester/Brockworth GDAs, with Bristol's AAOR at Lansdown, just over the border towards Bath, completed in 1954, and Gloucester's at Ullenwood, south of Cheltenham. Constructed of reinforced concrete these windowless bunkers had two floors of office, technical and domestic accommodation wrapped around a double-height operations room overlooked by balconies. Built at enormous cost, they ceased to perform their intended function within two or three years as the rotor system's AA gun element was shut down in 1955 when the future was seen to lie in surface-to-air missiles (SAM). Although SAMs would only be deployed down the eastern side of the country, they had their origin in the Southwest. The Army's Red Shoes missile, developed in Luton and issued as Thunderbird in 1959, was powered by Bristol Aerojet Gosling motors. The RAF's Red Duster project began development by BAC in 1947 and appeared as Bloodhound in 1958, also being deployed in West Germany.

The Royal Observer Corps (ROC)

The end of the war saw ROC No. 24 Group Gloucester continuing until 1953 when the group was disbanded and its posts redistributed, but its former HQ in Northgate Mansions remained as a Secondary Training Centre until 1976. Much of the 24 Group area was allocated to ROC No. 23 Group, reorganised as No. 12 Group, Bristol, which lasted throughout the period from 1953–1991. Its HQ in Kings Square Avenue was in operation from 1947–58, and then removed to protected accommodation in the only recently vacated AAOR at Lansdown, outside Bath, reporting to a Centralised Combined Filter Plot at RAF Box in the RAF Southern Sector. Expectations of a subsequent move to Hereford were overtaken by the abrupt stand-down of the ROC in 1991, by which time there remained just seven Gloucestershire posts in ROC No. 12 Group, although two new aircraft spotting posts had been added to the network in 1952 at Little Sodbury and Hucclecote. By the 1950s, although a few brick-built posts had appeared resulting from local initiative, the majority of the 1,500 ROC posts were still timber sheds or rooftops. Orlit, a firm specialising in buildings constructed from prefabricated concrete panels, was commissioned to produce a standard aircraft spotting post. This was a box, 10 feet (3 metres) long and 6 feet 8 inches (2 metres) wide and slightly lower than that in height. Half of it was roofed to protect the instruments, while the other, open, section allowed for visual spotting. The Orlit 'A' version stood at ground level and the 'B' version was raised on four concrete pillars 6 feet (1.8 metres) in height and accessed by integral steps. Five Gloucestershire posts received Orlit 'A's, including Hill and Dursley, while one Orlit 'B', still standing though currently obscured by foliage, was erected at Rodmarton. However, as ever faster aircraft appeared, it became increasingly difficult for spotters to make any meaningful contribution to the early warning process.

With perceptions of a nuclear threat from Russia's newly developed atomic weapons increasingly causing alarm, the United Kingdom Warning and Monitoring Organisation (UKWMO) was set up to provide feedback on any nuclear strike that might take place. As an organised and disciplined body spread evenly across the country and, furthermore, looking for a new role, the ROC was perfectly positioned to take on this new responsibility. It was soon realised that their current posts were far too exposed to do the job, so Orlit was again commissioned to design and supply an underground post. This was a larger concrete box measuring 20 by 8 by 8 feet (6 by 2.5 by 2.5 metres), sunk 15 feet (4.5 metres) underground, and accessed by a hatch and a vertical ladder. Inside was room for a crew of three with rudimentary sanitation, bunks, and power supplied by a car battery. On the surface only the concrete cube holding the entry hatch, a ventilator and the tops of the Bomb Power Indicator and Fixed Survey Meter Probe were visible, all painted green. The key instrument was the pin-hole camera mounted outside which would record the flash of a bomb-blast and give some indication of its position, Ground Zero, and its intensity. Available information would be telephoned or radioed into Group HQ for transmission to the SOC. These underground posts came on stream in the early 1960s and, although a large number were stood down in 1968, many were kept in service until 1991. Locations which were ideal as aircraft spotting posts might be less suitable for underground monitoring posts, so some relocations were required. The Almondsbury post was moved to Thornbury, and it was necessary to move seven other posts shorter distances. Entirely new underground posts were established next to a pillbox on the top of Nut Hill at Upton St Leonards, and on Stoke Orchard airfield. The Stroud post at Haresfield, relocated from Woodchester in 1961, closed in 1968 but was reopened in 1970 when the post on Nut Hill was closed. An example of one of No. 12 Group's trios in 1968 is made up of Stroud (12/J1), Leighterton (12/J2), and Dursley (12/J3). All these posts were crewed by unpaid volunteers, conscientiously training for an event they dreaded. Lansdown AAOR had been easy to adapt for a new role as ROC Group Control in 1959. The double-height operations room, overlooked by its first floor, newly glazed balcony, was standard for ROC controls. Ranged around this central hub were offices, canteens, GPO and radio rooms, male and female dormitories, ablutions, and lecture room. Whether those men and women, many of them volunteers, selected to staff this facility would leave their families outside to die instantly or more slowly remains a question which, fortunately, never needed to be answered.

The Royal Observer Corps post on Nut Hill overlooking Brockworth with a pillbox in the background.

RAF and USAF Operations and Airfields in the Post-War Period

Most of the county's airfields had been stood down by 1955, but some provided storage sites for the Quedgeley depots. South Cerney became the Central Flying School from 1952 and continued in operation until 1971, being outlasted by Aston Down which remained in operation as a RLG until 1976. Glosters used Moreton Valence for test flights of Meteors and Javelins until 1955 and kept Brockworth until 1960. The cramped location of Whitchurch saw flying ending in 1955 with Lulsgate Bottom in Somerset becoming Bristol's replacement airport. Staverton, however, enjoyed a future as a busy civil airport with a post-war ATC and a hangar, heightened in 2012, variously described as a T2 or a Bellman and so much modified as to be recognizable as neither.

After the war, Filton expected to have a place in RAF Fighter Command. To accommodate fast jets, one 4,500-foot-long (1,385-metre-long) runway was extended to 8,150 feet (2,510 metres) by demolishing an entire village. Its width was increased to 300 feet (92 metres) and a large turning circle was laid at each end. The other, shorter 4,000-foot-long (1,230-metre-long) runway enabled fighters to continue operating while the work was being carried out. Initially, No. 501 Squadron RAuxAF flew Spitfires but was upgraded to de Havilland Vampires in 1948. After 501 Squadron's disbandment, only the University Air Squadron, reformed in 1950, remained a regular user of the RAF part of the airfield. Filton's enormous Brabazon Hangar was erected after the war as an assembly building for the proposed new eight-engined eponymous airliner expected to be flown by BOAC. It consists of three bays each 331 feet (102 metres) wide and 65 feet (20 metres) high. The centre bay is 420 feet (130 metres) and the two side ones are 270 feet (83 metres) deep. The hangar was completed by September 1949 to coincide with the Brabazon's first flight. In the end, neither BOAC nor BEA wanted the aircraft, and the two prototypes were scrapped. The Brabazon Hangar was subsequently used to build thirty Bristol Britannia airliners and then Concorde. Although Filton was predominantly a civil establishment, BAC also used the facilities for the development of military aircraft when English Electric Lightning trainers were completed there in the 1960s.

In 1962, Filton's military role was revived when it was selected as one of twenty-two dispersal airfields equipped with a pair of Operational Readiness Platforms (ORP) linked to the main runway by a dedicated taxiway, and all other necessary facilities for the operation of a brace of Vulcan bombers armed with atomic weapons and to be stationed there in times of international tension. Aeronautical development and production continued at Filton up to 2012 when BAe handed over to Airbus. The single First World War GS hangar was used into the 1990s for missile development work. Despite flying activities being confined to emergency service helicopters, the site remains focused on aviation and Aerospace Bristol houses Concorde in the renovated triple GS hangar. Airbus has refurbished both Filton House and Pegasus House, two Grade II listed buildings within its office complex. Parts of the former airfield are currently (2021) being developed for housing. Alveston Manor, now a hotel, was bought in 1960 by Bristol Siddeley as a conference centre and, on their amalgamation, was taken over by Rolls-Royce and used until 2017.

In 1946, the switch from troop-carrying Halifaxes to the Hastings saw Fairford temporarily abandoned, but joint RAF and USAF use resumed in 1948. A base for USAF Strategic Air Command's (SAC) Convair B-36 Peacemaker heavy bombers was established and a 10,000-foot-long (3,000-metre-long) runway was laid for their use in 1953. Wimpey constructed a new hangar and built sixty-seven married quarters during the 1950s. Between 1964 and 1967, it hosted the RAF Central Flying School with the Red Arrows and then moved across to RAF Air Support Command for use by Hercules C1s. Later, the base was used by USAF B-47 Stratojet aircraft. For a ten-year period from 1969, Fairford provided a test centre for Concorde. From 1979, a squadron of KC-97 and 135 Stratotankers operated from Fairford, supporting the operation against Libya in 1986, and accompanying USAF B-52s during both Iraq wars in 1991 and 2003, and the NATO offensive against Serbia in 1999. More recently, B-2 Spirit Stealth bombers, accommodated in secure hangarage, have taken up residence. In 2000, the base received a $100 million upgrade, carried out by Alfred McAlpine Engineering Ltd, which included runway and apron resurfacing, and a tripling of fuel storage capacity to 9 million gallons (40.5 million litres). Two new Luria hangars were built measuring 320 by 160 feet (97.5 by 49 metres) and capable of holding the Stratojets. Reopening in 2002, it has been one of only a few emergency landing grounds available to NASA, its runway equipping Fairford to function as a stand-by airfield for everything from heavy bombers to space shuttles. USAF bombers currently (2021) use Fairford as a base for NATO exercises. Developments scheduled to commence in 2021 will see an increase in use by visiting aircraft, new buildings including a maintenance hangar for B-52 bombers, and a concomitant increase in base personnel.

From 1982, Little Rissington became a contingency hospital, one of three established by USAFE. One of its 'D'-type hangars was set up as a hospital ward with the potential to accommodate 1,000–1,500 beds, while the other storage hangars were earmarked for a similar transition. One 'C'-type hangar was a bed store; another provided storage for ambulances along with the equipment for field hospitals; and a third was fitted out with an automated racking and retrieval system for storing medical supplies. The three-storey officers' mess was to be a centre for the treatment of walking wounded. Although the airfield was cleared to accept Hercules and Galaxy aircraft and there had been a suggestion that the runway might be extended to take even larger aircraft, it was anticipated that casualties would arrive by USAF or *Luftwaffe* helicopter. The hospital was activated in 1990 at the time of the first Iraq war. Unfortunately, the suggestion that an enormous underground hospital lies under the airfield is a myth.

RAF Little Rissington, the 1930s officers' mess which constituted the walking wounded element of the USAF contingency hospital of the 1980s.

Other Aspects of Defence

The Royal Air Force

RAF Innsworth had become indelibly associated with the WRAF and continued to carry out training after the war's end. It housed No. 2 RAF School of Cookery and two Schools of Recruit Training along with a variety of administrative functions including the RAF Records Office from nearby Barnwood and RAF Personnel and Training Command. From 1997 until 2008, it was home to the joint-service Armed Forces Personnel Administration Agency. The logistics storage and maintenance facility at the RAF Quedgeley complex, which was using the RAF airfields at Moreton Valence, Babdown Farm, and Bibury in 1947, added further storage on its main sites, and by the time it closed, it covered 2.25 square miles (575 ha), continuing to operate until 1995. There are around twenty ATC units spread around the area in the major conurbations and in other centres including Quedgeley, Lydney, Westbury-on-Trym, Thornbury, Tewkesbury, Newent, and Innsworth.

The Royal Navy

HMS *Flying Fox* was retained until 1951 when she was replaced as the RNVR drill ship but remained on station until 1973. Her replacement, HMS *Locust*, a gunboat launched in 1939, served until 1968, when the Severn Division RNR moved into the Glosters' Bedminster drill hall retaining their HMS *Flying Fox* appellation. Dorset House in Litfield Place, close to the Clifton suspension bridge, became a Royal Marine Reserve base around 1963, utilising a pair of 1830s villas, with a drill hall being added by 1976. The Horfield Common ARC hosts the Filton Sea Cadet Unit, one of seven in the county, along with one of Royal Marines cadets.

Military Communications

After Bletchley Park closed in 1946, following a short stay at Eastcote in north London, in 1952, GCHQ moved to two sites, each of around 45 acres (18 ha) in Cheltenham: Oakley Farm and Benhall. Initially, the existing TOBs were occupied but new buildings, including the extensive C-Block, were completed at Oakley in 1954. GCHQ's research departments and the experimental radio station from Knockholt, Kent, were relocated to Benhall in 1952. The GCHQ Intercept School lay within the main establishment but later transferred to the Royal Signals Centre in Loughborough (Leicestershire). Fortunately, the large Cheltenham complex offered ample room for the serial expansions caused by the need to establish large computing facilities. In 1970, the Communications-Electronics Security Group moved in, causing further expansion. By the 1990s, it had become apparent that the fragmentation of the different departments was not conducive to ways of working in the new digital world, exacerbated by the decrease in staff, necessitated by the so-called peace dividend following the fall

Bedminster, the 1930s drill hall of the Glosters now serving as the Royal Naval Reserve Centre HMS *Flying Fox*.

Clifton, Litfield Place, Dorset House, the nineteenth-century house and post-war drill hall of the Royal Marines Reserve.

of the USSR. The new GCHQ building at Benhall, referred to colloquially as the 'Doughnut', opened in 2003.

RAF Pucklechurch in Bristol, the former balloon depot, became a joint services technical training depot, running ten-week courses on the inner workings of radio; on how to tune B40 to frequencies; and how to penetrate Russian wireless traffic through static. In 1962, 14 Signals Regiment moved into Robinswood Barracks. This unit had been formed in 1959 to replace the UK Commonwealth Communications Army Network (COMCAN) Signals Regiment which had relied on static high-powered machinery and enormous aerial arrays to connect London, Nairobi, Melbourne, Hong Kong, and Singapore. Its successor was involved in the trials of SKYNET, the communications satellite whose main user was GCHQ in Cheltenham. The regiment moved to Worcester in 1968.

The Birdlip radio station provided air traffic communications links to Europe and the Mediterranean from 1946 and the Barnwood OAMCU became an air traffic control centre for the Southwest region in 1947, this function being transferred to Uxbridge three years later. Birdlip remained part of the North Atlantic HF R/T network (NARTEL) until 1965 when its function was taken over by Ballygirreen, County Clare. Birdlip continued working as a receiving station with its associated transmitter station a few miles down the A417 at Winstone. Once consisting of two fields of aerials, most had disappeared by 1989, but two remain at Shab Hill and a further three at the Elkstone site at Winstone, modified in 2000. A proposed development of the site in 2006 as a telecommunications testing facility, allegedly associated with GCHQ, failed to proceed when Cotswold District Council felt that the construction of two lattice towers, one of them bearing a large radome, were incompatible with the site's location in an Area of Outstanding Natural Beauty. However, a new site at the Edgehills Radio Station at Plump Hill, Mitcheldean went ahead after an application in 2008.

In 1978, the Signals facility housed in a hardened bunker at RAF Boddington became No. 9 Signals Unit under RAF Support Command Signals HQ and RAF Strike Command HQ, 1978–2008, their respective networks coming under Communications Control Defence Network (CDCN). In 1998, operational command transferred to the newly formed Defence Communications Services Agency (DCSA) becoming DCSA 9 SU Boddington. The unit then became the primary message agent site for the UK as part of the Automated Messaging Systems Communications Equipment Replacement Programme (AMSCERP) providing messaging systems for the RAF and the British Army in static sites across the world. An upgrade took place in 2006. Once AMSCERP's function had been replaced by the Naval Shore Telegraph Network (NSTN), Boddington became redundant and closed at the end of 2007. No longer an RAF establishment but still under MOD control, it is currently (2021) operated by Information Systems and Services as ISS Boddington, part of Joint Forces Command.

In 1949, the GPO had commenced building a chain of experimental relay stations for the use of BBC TV and, later, ITA. One of these first six stations

Winstone, Birdlip Receiver station's transmitter partner.

was scheduled for Wotton-under-Edge and the steel lattice tower standing to the north-east of the village may represent this early installation. By the early 1950s, a new core communications network known as 'Backbone' was under construction. This was a chain of inter-visible microwave towers designed, in part, as an element of the military early warning system, and the later tower to the east of Wotton-under-Edge was justified at the planning enquiry in 1963 on wholly defence grounds. It is constructed of reinforced concrete and stands 250 feet (76 metres) high. It connects to Purdown to the Hawthorn terminal at Five Ways and, eastwards along the prime Backbone line, to Sparsholt Firs. In 1956, Backbone had gained spurs which brought defence facilities such as Boddington, Birdlip, and GCHQ Cheltenham into the network. Although the system was designed to obviate disruption to underground cabling in cities destroyed in a nuclear attack, a tower was nevertheless built in Bristol at Purdown. This was linked to Wotton-under-Edge and also to a further spur leading to a tower at Cinderford which would, in turn, link with Birmingham via Bredon Hill, north of Tewkesbury. It comes as a surprise that this defence-oriented system was subject to challenge by environmental groups, critical of the aesthetic inadequacies of the

Above left: Wotton-under-Edge, the microwave tower, part of the Backbone network.

Above right: Purdown, the microwave tower, unusually located in an urban area.

lattice towers. This objection may explain why the very visible Purdown, added to the network in 1970, is a concrete 'pencil' tower, 230 feet (70 metres) high, although an earlier tower of 1961 was the steel lattice model still standing next to its smart new neighbour in 1971.

In addition to the military radio masts and those of Backbone and its extensions, the Home Office established a chain of radio transmitters to enable the police to co-ordinate internal security in the event of a nuclear attack. Within sub-region 71, there were Home Office radio stations with masts on Cleeve Hill and at Cinderford.

Precautions against Nuclear Attack: Government and National Infrastructure

One of the principles of defence policy was a commitment to the improbable notion that life would continue following a nuclear attack. In the face of both scientific evidence and plain common sense, the government persisted in building and maintaining an ostensibly protected infrastructure to ensure continuity of authority, under the aegis of the Home Office which established a hierarchy of bunkers for regional and local government across the country. Gloucestershire came under region 7 whose war room, dating from the 1950s, is at Flowers Hill, Brislington, off the Bath Road. Typical of those buildings constructed in the dozen regions, it consists of a massive, two-storey block which stands in an estate of TOBs. Like the contemporary AAORs, there is a double-height map room with the customary balcony surrounded by domestic facilities on two floors. Standout features are the trio of ventilation ducts on the 7-foot-thick (2.15-metre-thick) reinforced concrete roof; the 5-foot-thick (1.5-metre-thick) walls; the solid blast doors; and the internal airlocks. A single-storey annexe contained the machinery for heating, ventilation, and air conditioning. By 1962, the devolution policy was still being maintained but the existing regime of regional centralisation was modified, with most regions being given two sub-regional centres. Region 7 was split, with one centre at Bolt Head in Devon occupying a former Rotor GCI radar bunker, and the former AAOR at Ullenwood, south of Cheltenham, housing the other, being classified as a Regional Seat of Government (RSG) and retaining this function until 1985 when a new RGHQ at Chilmark, near Salisbury, took over.

Below the RSGs, the next level was the county emergency control. The basement of Shire Hall, Gloucester, filled this role until 1985 when Ullenwood became available as the county's emergency HQ, and the short-lived Avon County Council, although unhappy with the level of rent charged, leased the Brislington war room from the Home Office. In Bristol, there was a bunker underneath the Council House on College Green and four CD control centres, at Banwell Close, Bedminster, in Avonmouth, in Henleaze, and in Speedwell.

Brislington, the Cold War regional war room for the south-west of England which later served as Bristol's Civil Defence control centre, containing a double-height map room with domestic facilities on two floors. Three ventilation ducts stand on the 7-foot-thick (2.15-metre-thick) reinforced concrete roof.

These were single storey concrete blockhouses built in 1954 and abandoned in 1968. South Gloucestershire District Council's offices in Thornbury, opened in 1986 and recently sold as a retirement home, included a small, hardened control bunker to the rear. Kingswood Civic Centre was scheduled as a further possible emergency centre but only to be fitted out if found to be necessary. Astonishingly, Stroud Council's purpose-built HQ in Ebley Mill, built as recently as 1989, was equipped with an above-ground bunker with concrete walls and slab roof, subsidised by the Thatcher government just one year before the end of the Cold War. Since 1994, the former AAOR and ROC Group HQ at Lansdown has served as a fire service training facility.

Bunkers were also provided in attempts to safeguard public utilities in the event of a nuclear strike. The water companies building protected control centres included Thames Water with eight such installations being built in the region in the 1980s, one of them at Donnington near Stow-on-the-Wold. Severn-Trent Water installed a bunker under an office block at their Staverton depot, completed in 1992. It benefits from the usual features of such a bunker with blast doors and air locks and special protection being afforded to the ventilation system to avoid it being compromised were the building above to collapse. The National Grid also maintained a network of Strategic Reserve Depots in hardened buildings. These stored all the equipment and plant needed to restore power in the event of

Barnsley, the National Grid emergency store which contained equipment for restoring electricity supplies following a nuclear attack.

Donnington, the protected Severn-Trent water authority's control centre.

a nuclear blackout. One of them was located at Barnsley, north of Cirencester, in the 1960s.

One particular governmental concern in the aftermath of a nuclear attack was the retention of freedom of movement for military and CD traffic, and the imposition of limits to the general public's ability to travel. To this end, most major roads, especially those which bypassed major conurbations, were designated as essential service routes. These included the M4, M5, M32, A40, A46, and A436. Civilian traffic would be barred from these roads and this policy would be enforced by armed TA troops. Additionally, the Military Road Route System ensured that communication between key points would be safeguarded in order that military traffic was free to access bases and military ports. Locally, these included Gloucester to Colchester (ELK) via the A40; Cirencester to the M1 (DOG) via the A429; Cardiff to the M5 (COW) via the A48, A449, and A40. Civilians would be free to use other roads but, with widespread damage to the transport infrastructure anticipated in the cities, no way of replenishing fuel, and little room for manoeuvre, minor roads would soon clog, and few people would make their escape to the countryside.

As had been the case during the Second World War, government departments including the Ministry of Agriculture Fisheries & Food (MAFF) maintained Buffer Depots containing food for distribution in any conditions of national emergency. These included depots at Badminton (362AS); Gorlands Road, Chipping Sodbury (363T); and Hawthorn (96QQ) in region 7, and Station Approach, Tewkesbury, in region 9. Selected schools were briefed to serve as first aid and community feeding stations.

The Aircraft Industry

In 1946, BAC offered a design for a long-range, high-speed bomber powered by four turbo jets. A mock-up of the swept-wing aircraft was constructed but suffered from aerodynamic failings and BAC withdrew to focus on the Britannia, eighty-three of which were built at Filton from 1952, some of them to fulfil RAF orders. The company also designed the Bristol Type 192 Belvedere as a twin-engined tandem-rotor military helicopter in service with the RAF from 1961 to 1969. It could carry nineteen troops and their equipment, or twelve stretcher cases. In 1958, BAC joined with other companies to form Bristol Aerojet (BAJ). Located at Banwell, just across the Somerset border, BAJ worked on a range of missiles including Blue Water and Black Arrow along with, from 1967, the development of Chevaline which would update the Polaris nuclear deterrent, allowing it to remain in service with the Royal Navy until 1998. When the Rotor AA system was stood down in 1955, BAJ was already working on casings and propellants for the new generation of SAMs. In 1966, the standalone Bristol

Aero Engines, already merged with Armstrong Siddeley, was purchased by Rolls-Royce, which continued to produce Bristol's Olympus aero engine. Rolls-Royce was nationalised in 1971, and BAC itself was nationalised as part of British Aerospace (BAe) in 1977. In 1984, VC10 airliners were being converted as tanker aircraft for the RAF at Filton. A further merger took place in 1999 between BAe and Marconi to form BAe Systems.

The Brockworth and Hucclecote aircraft factories of Gloster Aircraft Company operated until 1964 when they were taken over by British Nylon Spinners. Glosters' Bentham factory had continued in use as a development facility, following up the Meteor with the Javelin, but has now been demolished. Dowty Equipment Ltd acquired Rotol in 1960 and continued work at the Staverton site, producing propellers and other parts for, among other aircraft, the Hawker Hunter and the Lockheed C-130 Hercules. Following a severe fire at Staverton in 2015, the company moved into a new building on the Brockworth site in 2019, but its sites in Cheltenham and at Staverton, fronted by the magnificent Art Deco office building, are still in use. Sperry, one of whose plants remained at Stonehouse until 1969, was responsible for the inertial guidance system of Blue Streak, Britain's first attempt at building a ballistic missile, which was terminated in favour of an off-the-peg model from the USA.

Staverton, the Art Deco office building for Rotol Airscrews.

APPENDIX I

Prehistoric Hillforts

Entries in italics denote doubtful or rejected sites.

Alveston, Little Abbey Camp ST650887
Amberley Camp SO852014
Bagendon Hillfort SP018064
Batsford, Queen Victoria's Garden SP203329
Beckbury Camp SP064299
Bibury, Ablington Camp SP105074
(*Charlton Kings, Battledown Camp SO970220*)
(*Birdlip Camp SO924150*)
Blaise Castle Camp, ST559784
Blisbury Hill Camp ST656969
Brackenbury Camp ST747949
Buckland, Burhill Camp SP085363
Bury Hill Camp, Mangotsfield ST652791
Charlton Abbots Camp SP047243
Cheltenham, Battledown Camp SO970220
Chosen Hill or Churchdown Hill SO882187
Cleeve Common (part quarried) SO985255
Cleeve Hill, The Rings SO984266
Clifton Down Camp ST566733
Cold Ashton, Tog Hill ST733722
Colesbourne, Norbury Camp SO990149
Condicote, Eubury Camp SP157288
Crickley Hill SO928161
Cromhall, Bloody Acre Camp ST689915
Damery Old Castle Camp ST707944
Dean Camp, Coln St Aldwyn SP176087
Ditches Camp, Woodmancote SO996094
(*Dixton Hill camp, Alderton SO986305*)
Drakestone Camp, Stinchcombe ST736981
Dyke Camp, Harescombe SO835086
Dyrham Camp ST741767
Edgeworth, Juniper Hill SO931064
Elberton Camp ST608884
Hailes Wood Camp, Stanway SP056301

Haresfield Camp SO823090
Hazelwood Copse Camp ST858985
High Brotheridge Camp & Coopers Hill SO891139
Horton Camp ST764844
Icomb Camp SP204232
King's Weston Hill Camp, Henbury ST556781
Knole Park camp ST596833
Ladborough Camp SP165087
Lancaut promontory fort ST541966
Leckhampton Hill SO948184
Lydney Hillfort SO616027
Norbury Camp SP127156
North Dowdeswell SO999191
Nottingham Hill SO984284
Oldbury-on-Severn, The Toots ST611927
Oxenton Hill, The Knolls SO972314
Painswick Beacon SO869121
Pinbury Hillfort SO958053
Ranbury Ring, Ampney St Peter SP090009
Rockhampton, Camp Hill ST659927
Salmonsbury Camp, Bourton-on-the-Water SP175209
Salperton Camp, Hazleton SP066194
Shenberrow Hillfort SP080334
Sodbury Camp ST761826
Soudley Camp SO661105
(*South Dowdeswell SP004186*)
Stow Camp SP194260
Stowe camp SO564064
Sudeley, Roel camp SP047244
Symonds Yat Fort SO563157
Tetbury Camp ST891928
Thornbury, Camp Hill ST657927
Towbury Camp SO880370
Trewsbury Hill fort ST981998
Tytherington, The Castle ST664887
Uley Bury Camp ST785989
Welshbury Camp SO678156
Willersey Hill Camp SP120386
Winterbourne Down ST652791
Windrush Camp SP181123.

APPENDIX II

Castles, Moated Sites, and Strong Houses

The sources for Appendix 2 include the *Gatehouse* website; David Cathcart King's *Castellarium Anglicanum*; Charles Coulson's lists of Licences to Crenellate; and Barbara Rawes' Gloucestershire checklists published in *Glevensis* 11 and 12. Doubtful sites in italics.

Alderton, Dixton Hill SO986305 doubtful motte and bailey on extensive hilltop site
Arlingham, Wick Court SO736105, moated house with the base of a square, stone tower
Berkeley ST685990 Norman motte revetted in stone to form a shell-keep with projecting towers, inner and outer baileys with domestic ranges and gatehouses
Beverston ST862940 stone castle of 1229 with residential tower and gatehouse
Bledisloe SO683082 motte destroyed in the 1970s
Brimpsfield SO946127 small, low motte
Brimpsfield SO942127 a quare keep of 1200s; inner bailey with square towers and gatehouse; and outer bailey containing hall and chapel with further towers and gatehouse
Bristol ST592731 motte replaced by large rectangular stone keep with inner bailey; outer bailey with hall; curtain walls and gates with a postern accessed via a tunnel
Cirencester, location unknown but in existence in 1142
Coberley (Emmas Grove) SO934915 possible castle mound
Coberley SO965157 marked as castle by OS but mounds would all appear to be barrows
Cromhall ST692904, fortified house licensed in 1311
Culverden (SO835238), probably Willington Court in Sandhurst, where the remains of a thick-walled medieval tower survive within the later house which was probably originally moated; licensed in 1318

Dean, unidentified, possibly spurious site, suggested as Stowe Green (qv), Littledean Camp (qv) or Welshbury Camp (see Appendix I)

Dursley/Durslea, earthworks at Drakestone Point ST737980 may have been a beacon or lookout; foundations have been reported at an alternative site at ST755982

Dymock (Aylesmore Castle Tump) SO712293, motte mentioned 1148–54

English Bicknor SO581157, motte and bailey with church in outer enclosure, mentioned 1217

Gloucester SO832183 began as a motte which later became the castle's barbican, with a large stone keep, three inner wards and an outer one accessed through gatehouses and bridges over the wet moats

Great Howle Camp SO612202, suggested as analogous to the eleventh-century Littledean Camp and therefore a possible Norman ringwork

Hailes (Stanway) SP052302, earthworks of 1139–48 marked as a castle by OS and may be associated with the adjacent abbey

*Harescombe SO837104, phantom castle marked on 1885 OS Map; dismissed by work in 1970 and 2000 (*Glevensis 40, 2007*)*

Haresfield Castle SO810105, low motte near church

Hasfield Court SO827274, moated early Tudor house with gatehouse

Hewelsfield SO567021, doubtful motte SW of church

Hillesley ST768897, no traces remain of this possible earthwork castle

Icomb Place SP211225, semi-fortified house, remodelled in 1420, with two courts and an embattled gate-tower

Kempsford SU161964, a semi-fortified courtyard house with moat and gatehouse built on earlier earthworks but demolished in 1610

Lasborough Farm ST824941, a low, damaged motte, apparently with no bailey

Littledean SO677135, a strongly banked ringwork in use around the turn of the eleventh century

Lower Slaughter (St Mary's Church) SP166226, appears to have provided a secure temporary base for brigands during the Anarchy

Lydney (Little Camp Hill) SO617024, motte and bailey of the 1100s with foundations of a rectangular keep and a five-sided inner bailey with a gate-tower and a further square turret

Miserden SO944093, motte carrying foundations of a shell-keep, mentioned in 1146

Newington Bagpath (Kingscote) ST816947, small motte bearing possible foundations of a collapsed tower

Newnham-on-Severn SO689115, ringwork described as 'old' in 1240

Old Ley Court (Westbury-on-Severn) SO744179, a semi-fortified house, licensed in 1330

Olveston Court ST599871, late fifteenth-century house with, moat, embattled wall and gatehouse but any 'defences' were confined to only three sides of the enclosure

Painswick (Castle Godwyn) SO871116, a possible motte but lacking clear evidence

Ruardean SO619180, stone castle licensed in 1311 of which slight traces remain

St Briavels SO558045, imposing twin-towered late thirteenth-century gatehouse with attached hall and chapel within an earlier bailey; square keep, formerly adjoining bailey curtain, now reduced to foundations; used as hunting lodge, courthouse, prison, and arsenal

Soudley (East Dean) SO661106, possible ringwork cutting off the end of a promontory but is more likely to be prehistoric in origin

South Cerney SU047976, fragmentary remains of castle captured by Stephen in 1139

Standish Court SO801084, fourteenth-century semi-fortified house of abbots of Gloucester with gatehouse

Stowe Green (St Briavels) SO564064, ringwork

Sudeley SP032276, picturesque quadrangular castle of 1400–50 with robust Dungeon Tower, traceried windows in hall range, further towers and outer court

Taynton (Old Church), SO747228, a ringwork converted to a motte with multiple baileys

Taynton (Huntley) SO715211, ringwork of similar form to Littledean Camp (qv)

Tetbury ST890929, earthworks near church possible site of castle mentioned 1144

Tewkesbury (The Mythe Tute) SO888340, much damaged motte and bailey eroded by the river

Tewkesbury (Holm) SO891321, earthworks and residential buildings fortified with brattices in 1211

Tewkesbury (King John's Castle) SO889341, three- (formerly four-) storey rubble-walled staircase tower of abbot's lodging

Thornbury ST634908, impressive remains of Lord Stafford's unfinished palace, licensed in 1511, but adjoining the site of an earlier fortified manor-house. The towers are intended to impact as a martial statement, backed up by the courtyard of retainers' lodgings

Upper Slaughter SP156233, small, poorly preserved motte containing deep, stone-lined well

Wanswell Court SO690016, medieval house dating from 1256 with a hall house of 1450 as the core of a later building standing on a large, moated platform contrived by damming a stream which also created fishponds

Whitminster/Whitenhurst/Wheatenhurst SO760091, possibly the courtyard house licensed in 1347 but not moated and probably never seriously fortified

Winchcombe SP023281, a castle was erected by Roger of Hereford in 1144 near the church and possibly based on the burghal defences but was only short-lived

Withington SP030156, fortified house of the bishops of Worcester, licensed in 1271; now occupied by later buildings the medieval complex included a gatehouse mentioned in 1476

Yate Court ST712860, medieval moated, strong house licensed in 1299; only fragmentary ruins of an inner gate tower and part of a hall remain on site, but a gatehouse was removed and re-erected at Berkeley Castle in the twentieth century

APPENDIX III

Licences to Crenellate

Sources are Charles Coulson (pers com) and Philip Davis (Castles Study Group Journal 20, 2006–7)

1153: Robert Fitzharding of Berkeley Castle is recorded as being granted a charter to rebuild the castle
1229: Beverston Castle to Maurice de Gaunt
1271: Withington to Godfrey, bishop of Worcester
(1291: Little Compton to John, archbishop of York. The licence for this house, now in Warwickshire, may never have been implemented)
1299: Yate Court (*ST712860*) to John de Wylington
1311: Cromhall *(ST692904)* to William de Wauton
1311: Ruardean (*SO620178*) to Alexander de Bicknor
1318: Culverden (*SO835238*), to Henry Wylington; probably Willington Court in Sandhurst
1330: Westbury Old Ley Court (*SO744179*) to Robert de Sapy
1334: Boddington (*SO895257*) to John de Bures & Hawise his wife. Their house was moated but otherwise unlikely to have had serious defensive features.
1347: Whitminster/Whitenhurst/Wheatenhurst to Humphrey de Bohun, duke of Gloucester
1373: Winchcombe Abbey to the abbot, probably relating to a precinct wall and gatehouse
1458: Sudeley Castle to Ralph Boteler
1510: Thornbury Castle to Edward Stafford, third duke of Buckingham

APPENDIX IV

Airfields and other Aviation-Related Sites

Aston Down opened in 1938 as an ASU run by 20MU; a fighter pilot group pool, 5 OTU from 1939 then 52 OTU from 1941; ATA ferry pool until 1945; Fighter Leaders' School in 1943 from Charmy Down; 3 Tactical Exercise Unit practising rocket-firing; then 55 OTU; 83 GTS in 1946 from Moreton Valence; RLG until 1976; currently (2022) a trading estate.

Babdown Farm started as a RLG for 9 FTS then 9(P)AFU at Hullavington (Wiltshire); from 1942, No.3 Flying Instructors' School and Beam Approach Training; training ceased in May 1945 and the airfield became a sub-site for the Quedgeley depots until January 1948; currently (2022) a trading estate.

Barnsley Park (No. 22 SLG) came into service in July 1941 under No. 6 MU at Brize Norton.

Barnsley Wold was a proposed USAAF Transport airfield from the 1943 Review which never materialised.

Bibury opened in April 1939 as a relief landing ground for 3SFTS at South Cerney but remained closed until July 1940 when night-flying resumed with Hurricanes and Spitfires of RAF No. 10 Group. From 1941, it hosted training units: 3SFTS, 3 (P)AFU and 1539 BAT Flight. After Christmas 1944, the airfield was taken over by No. 7 MU at Quedgeley for storage.

Bristol, Beggars Bush was used for liaison flights by light aircraft serving the US 1st Army HQ at Clifton College.

Brockworth/Hucclecote began as an aircraft testing site for Martyn's of Cheltenham in 1915 and was developed in 1918 as an Aircraft Acceptance Park (AAP). The site was bought by Gloster Aircraft Co. in 1926 and used by them until 1960. Brockworth represented the main Glosters factory and Hucclecote was added as an Air Ministry shadow factory in 1939.

Chedworth opened in 1942 as a satellite of Aston Down; Fighter Leaders School in 1943 as an offshoot of 52 OTU carrying out tactical training; it closed in 1945.

Cheltenham, Marlborough House (No. 109 Winchcombe Street) was No. 9 School of Military Aeronautics in the First World War.

Cheltenham, Lilleybrook Golf-course was used by US Service of Supply (SOS) liaison aircraft.

Cheltenham, Western Road: AM Works Department No. 9 Works Area.

Cirencester Park was HQ No. 21 RFC Training Wing which arrived from Filton in 1916.

Cirencester was a temporary Landing Ground in 1918 prior to Minchinhampton opening.

Cirencester (*SP010054*) was an Automobile Association Landing ground in the 1930s.

Cirencester, Royal College of Agriculture: HQ RAF Technical Training Command 1939–45.

Clifton, Elmdale Road was No.6 School of Military Aeronautics in the First World War.

Dowdeswell Court is a large Georgian Mansion requisitioned by the RAF during the Second World War as, initially, a college for RAF apprentices, and then from 1945 until 1961 as the RAF Chaplains' School which had been established at Magdalene College, Cambridge in 1943.

Down Ampney opened in June 1943 as a RAF Transport Command airfield near Cirencester. Gliders operated from here for Operations Tonga, Mallard and Market 1&2 using Horsas towed by Douglas Dakotas. After Market, Down Ampney, along with Fairford, was host to elements of No. 22 HGCU helping to train glider pilots.

Down Farm/Westonbirt (No. 23 SLG) was in use from May 1941 under Hullavington.

Fairford was intended for the USAAF, and construction was begun in 1943 by McAlpine's, but on completion it went to the RAF. Training focused on airborne operations for glider-borne landing and paratroop drops. Aircraft from Fairford flew on D-Day and in the Arnhem operation as well as actions with SOE and the SAS. Fairford was temporarily abandoned in 1946 but by 1950 the airfield had been handed over to the USAF and a new runway was laid in 1953 enabling it to become a forward base for the Convair B-36 Peacemaker. It then provided a base for USAF B-47 Stratojets of the SAC. Between 1964 and 1967, it hosted the RAF CFS with the Red Arrows then joined RAF Air Support Command flying Hercules C1s. From 1978, Boeing K-135 Stratotankers were based here and were involved in the 1986 operation against Libya and together with USAF B-52s flew during both Iraq wars in 1991 and 2003 and the NATO offensive against Serbia in 1999. From 1969–77, the 9,993-foot-long (3,046-metre-long) runway allowed use of Fairford as a test centre for Concorde and continues to equip Fairford as a stand-by airfield for everything from heavy bombers to space shuttles.

Filton began under the British & Colonial Aircraft Company commencing aircraft production in 1910. In 1915, it became a RFC Training Station then an AAP. Aircraft production continued throughout the interwar period and a Reserve Flying School operated from 1923. New aircraft types were developed during the 1930s and production was stepped up during the Second World War when service as an operational RAF fighter station also resumed when Filton was Sector HQ and Operations Room for No. 10 Group Fighter Command during 1940, before moving to Colerne in Wiltshire. No. 2 E&RFTS was forced to move to Staverton and the fighters moved to Bibury and Colerne. At the end of 1943, fighters and transport aircraft of the IXth USAAF Tactical Air Force arrived. In 1962, Filton was one of twenty-two dispersal airfields for the operation of two Vulcan bombers armed with atomic weapons and stationed there in times of international tension. In 1977, BAC merged with the Hawker-Siddeley Group to become British Aerospace and then in 1999 with Marconi to form BAE Systems. BAE and Rolls-Royce, nationalised in 1971, continued aeronautical development and production at Filton up to 2012 when BAE handed over to Airbus.

Gloucester, Barnwood was the Overseas/Oceanic Aircraft Control on Eastern Avenue, concerned from September 1940 with foreign aircraft movements and liaising with the W/T station at Birdlip, under RAF Ferry Command No. 44 Group from 1941. The RAF Record Office, in the building opposite, which had absorbed the Polish Air Force record office in 1946, transferred from Barnwood to Innsworth in 1951. In 1960, the traffic control functions moved from Eastern Avenue to Uxbridge.

Gloucester's RAF Purgatory Unit handled aircraft prior to their completion and issue during the Second World War.

Innsworth opened in 1940 as No. 7 School of Technical Training for engine and airframe fitters and mechanics. No. 2 WAAF Depot opened in December 1941 and training of WAAF personnel became the station's predominant activity, alongside other administrative functions mainly relating to personnel issues. From 1994, it was HQ of RAF Personnel and Training Command. The depot closed in 2008 and was handed over as a new Imjin Barracks, home to the NATO Allied Rapid Reaction Corps.

Leighterton opened in 1918 as an Australian Flying Corps FTS until early 1920. Lasborough Park Mansion housed the officers' mess and Lasborough Manor House provided junior officers' billets. An Automobile Association Landing Ground (ST830926) operated in the 1930s on the site of the First World War airfield, and glider training was still taking place in 1939.

Little Rissington opened in 1938 as No. 6 SFTS, 1939–40, using Kidlington (Oxfordshire) as a landing ground. It was home to the CFS, 1946–76, then served as Imjin Barracks, vacated by the Royal Irish Rangers by 1982. It then became back-up to USAFE at Fairford as a contingency hospital until 1992. Since closure in 1994 much of the complex has been developed as the village of Upper Rissington only retaining glider training for cadets.

Long Newnton was initially earmarked as a 'Q' decoy site for RAF Kemble (Wiltshire) but construction was underway in early 1941 as a relief landing ground for South Cerney which used it as a dispersal area. It was then used for glider trials by No. 2 GTS from Kidlington (Oxfordshire).

Minchinhampton was AFC No. 1 Training Squadron Station from 1917 until 1919. It was then abandoned and returned to agriculture, only to be requisitioned again in 1937 as RAF Aston Down (qv).

Moreton-in-Marsh was under construction in September 1940 as an OTU. The airfield opened in 1941 flying Wellingtons and Ansons. Squadrons took part in raids from the spring of 1942, and six Ansons were destroyed on the ground when the station was bombed in the July. In November 1946, No. 21 OTU was disbanded, but in 1952, in response to the demands of the Korean War, pilot training resumed under No. 1 FTS flying mainly Harvards and Prentices. Very little construction was carried out and flying ceased in early 1955. The RAF Fire Service and the Home Office continued to use the site and it became the Fire Service Training School which it remains.

Moreton Valence opened as a landing ground for Staverton's observer training units. The Pilot Refresher Training Unit was formed in 1942 but soon moved to Scotland leaving the observer unit in sole occupation. MAP set up flight test facility for the Gloster Meteor, using an extended runway by 1943. After 1946, the airfield with enhanced facilities was run by Glosters for final assembly and testing of the Meteor and then the Javelin which flew from November 1951.

Northleach opened in 1942 as a RLG for Stoke Orchard but both airfields suffered from bad surfaces and activity had been moved away by the end of 1944.

Overley (No. 14 SLG) operated from the beginning of 1942 under the control of No. 20 MU at Aston Down, and accommodated a wide range of aircraft, including four-engined types.

Pucklechurch (No. 11 Balloon Depot) opened in August 1939 as the base for three RAuxAF squadrons manning the Bristol Balloon Barrage. It consisted of three large balloon sheds and ancillary and domestic buildings. Two of the sheds remain (2017) as part of a trading estate adjacent to HMP Ashfield.

Quedgeley RAF supply depots opened in 1939 and operated until 1997. They occupied seven extensive, discrete sites with a large HQ building. Many of the sites have been redeveloped.

Rendcomb opened in 1916 as No. 38 Training Squadron Station and expanded as No. 5 TDS in 1918 but was then run down until it was closed in 1919. Unused in the Second World War, it is now a private airfield using reconstructed First World War buildings.

Shipton Moyne was considered by the Air Ministry as a possible location for a USAAF Transport Command airfield, but the plan was put on hold and never taken up.

South Cerney opened in 1937 as a purpose-built training airfield for No. 3 SFTS, intended for Short Service Commission pilots, using Wanborough (Swindon) as a landing ground. HQ No. 23 Group, RAF Training Command, arrived in 1939, staying until 1946. Training continued throughout the war with 3 (P) AFU and No. 7 Flying Instructors School until 1945 and beyond. In 1948 No. 2 FTS formed, evolving in 1952 into the CFS (Basic) Squadron. In 1971, the station was transferred to the Army as Duke of Gloucester Barracks, currently (2021) a base for 29 Logistic Regiment.

Southrop opened in August 1940 as a RLG for No. 2 FTS, later serving units from South Cerney. It closed in January 1945.

Staverton began life as an AA Landing Ground (*SO880223*) one mile north of Staverton airfield which was started as a municipal aerodrome in 1936. The Air Ministry carried out improvements in order to use it for training activities. No. 2 EFTS moved in from Filton in 1940 becoming No. 6 Air Observer Navigation School. The airfield also provided facilities for Rotol's propeller testing, using nearly ninety marques of aircraft. Follands was evacuated to Cheltenham so used Staverton for trials, sharing a hangar with Flight Refuelling Ltd. By late 1944, most flying was carried out by these commercial firms. Official use of the airfield ceased in 1950. Since then, various small-scale operations have continued until now, Gloucestershire Airport is busy with mainly light aircraft.

Stoke Orchard opened when No. 10 EFTS arrived from Weston-super-Mare in September 1941. This was followed in July 1942 when No. 3 Glider Training School was formed with Hotspur gliders. This operated, using Northleach as a relief airfield, until it was moved to Exeter in January 1945, leaving the airfield to close at the end of the year. Glosters ran a dispersal factory on the north side of the airfield assembling Typhoons.

Stonehouse, Wycliffe School: Air Ministry Meteorological Office and Accounts Branch.

Tetbury, Westonbirt School and Hare & Hounds Hotel: Air Ministry Meteorological Office.

Tewkesbury (Overbury) was an Automobile Association Landing Ground (*SO950353*) lying in Worcestershire, but its description locates it on the county boundary.

Tresham was considered by the Air Ministry as a possible location for a USAAF Transport Command airfield, but the plan was never implemented.

Whitchurch (Bristol) opened in 1930 as the airport for Bristol, only the third civil airport in the country. In 1939, the airport was requisitioned by the Air Ministry's Department of Civil Aviation, flying throughout the war to destinations including Lisbon. After the war, flying continued but owing to the lack of space for expansion, Lulsgate Bottom, over the county boundary in Somerset, was bought from the Ministry of Aviation in 1955 as Bristol's new airport.

Westonbirt School and **Arboretum** were used for aircraft storage from 1942.

Windrush opened at the height of the Battle of Britain in August 1940 as a RLG for No. 6 SFTS and then for No. 6 (P)AFU from Little Rissington. It closed down in July 1945.

Yate opened as the Western Aircraft Repair Depot (ARD) in 1917, employing nearly 2,000 personnel on an extensive site. Damaged aircraft were restored and engines reconditioned. The site closed in 1920 but aircraft production at the Parnalls factory would continue into the Second World War, with large numbers of powered gun turrets for military aircraft being produced.

APPENDIX V

Drill Halls, TA, and Army Reserve Centres

Notes: X denotes demolished, and * denotes extant in 2021; numbers in brackets show data from a survey of TACs carried out in 1958 recording numbers of drill halls in operation in 1910 and 1933. This 1958 record, unfortunately, does not include Bristol.

Avonmouth

X Ermine Way: the drill hall was built in the access to a clay pit in Barrow Hills Crescent, on the site of the Shirehampton Remount Depot and has been demolished since 1990. Nearby Barracks Lane commemorates the troops who manned the early nineteenth-century battery marked at the old mouth of the Avon and the range beyond the end of the lane.

Bedminster

*Winterstoke Road (A3029) drill hall, 1940; currently HMS *Flying Fox*, RNR.

Berkeley

(One in 1910)
Possibly the gymnasium behind 9 Salter Street later used by Boy Scouts etc.

Bristol

X Ashton Gate Exhibition Centre of 1914 was purchased by the War Department for use as a drill hall, canteens, barracks, mess rooms, rifle range, etc. for the Glosters' 12th (Service) Bn.

*Brislington, Bath Road: cadet hut.
*Clifton College, HQ 2nd Corps Gloucestershire Volunteer Engineers 1880.
*Clifton, 2 Beaufort Road: former CTA offices.
*Clifton, Litfield Place, Dorset House: originally a pair of 1830s villas with a drill hall, erected by 1976, in use by the Royal Marines Reserve since *c.* 1963.
X Colston Fort, Montague Place, Kingsdown; house of 1823, in use by HQ, A, B & C Sections RAMC Divisional Field Ambulances, 1914; RGH in 1939. Destroyed by bombs.
*Eastville, 'Glenavon' Daisy Road: HQ 'D' Squadron, RGH, 1902.
Broadmead, Carriage Company's Yard: Accoutrement Depot, 'D' Squadron, RGH, 1902.
*Headley Park, Headley Lane: former TAC and depot.
X Horfield Barracks, drill station for 'D' Squadron, RGH, 1914.
*Horfield Army Reserve Centre, opened in 1930s; 44 RTR and 266 HAA Regiment, RASC, RE, RAMC and Royal Signals units in 1947; in curent use as HQ of Military Intelligence units and 39 Signal Regiment.
*Kingsdown, 2nd Southern General Hospital, RAMC, based in Royal Infirmary, Upper Maudlin Street, recorded 1903, in use 1914. In 1947, 9th (Southern) General Hospital, RAMC (TA).
*57 Old Market Street, drill hall for 4Bn Glosters, opened 1915 as substitute for Park Row. Converted to apartments.
32 Park Row, HQ 1st & 2nd Field Coys and HQ & No. 1 Section Signals Coy. South Midland Divisional REs, 1914. Built in 1910 as the Coliseum skating rink, part of the long façade of Italianate arches is incorporated into 'University Gate', the university's engineering faculty opened in 1996.
*Purdown, Tackley Road: TA REME workshops and cadet hut.
X Queens Road, Clifton drill halls in use 1903 absorbed into museum/university buildings.
*St George, Cloud Hills Road/Cherry Orchard Lane, Baptist Church Hall (built 1853) used as a school in 1882 but in use as a drill hall by 1902 for 'F' Coy, 1st Vol. Bn Gloucestershire Regiment, and for 4th Bn Glosters in 1914; rented by Baptist Church from 1911; purchased 1912; Sunday School from 1932. Currently (2021) Supported Living centre.
X St Michael's Hill, house, after 1883, used as drill hall with drill ground to rear, now under St Michael's Hospital.
*Speedwell, Whitefield Road, TAC; converted to flats.
X Trinity Street, College Green, drill hall in use by 1902 and in 1914, by Bristol Volunteer Engineers; demolished.
*Whiteladies Road, Artillery Volunteers' drill hall, in use by 1903; HQ, 1st & 2nd Batteries and Ammunition Column, 1st South Midland Brigade RFA, 1914. In use (2021) as ARC by RA and RE units plus Bristol UOTC.

Cheltenham

(Two in 1910; one in 1933)

X North Street drill hall, opened 1906, designed by HT Rainger; 'E' & 'F' Coys, 5Bn 1908; 'A' Squadron, RGH, 1914; HQ 523 LAA Regiment (TA) in 1947; now site of Primark store.

*Swindon Road, New Rutland Court, RE Volunteer drill hall in use 1903, now a garage.

*2 College Lawn, HQ 'H' Squadron, 2nd RGH, 1930s.

*Arle Road, ARC (similar building to Cinderford, 1957) currently in use by REME, Army Medical Services and a company of 6th Battalion, The Rifles.

Chipping Campden

(Two in 1910)

*The Square, Bank building was the armoury of 'H' Coy, 5th Battalion, Glosters, 1914.

Cinderford

X Station Street, drill hall on the site of the present Telephone Exchange. In 1914, 'E' Company of the 1st Battalion, Herefordshire Regiment (TF) had a drill station in Littledean, adjoining Cinderford (possibly at the Royal Forester PH).

*Valley Road, drill hall, 1957; Herefordshire Regiment badge over door. 'A' Company, 1st Battalion, Herefordshire Regiment by 1962; in use by ACF, 2021

Cirencester

(Two in 1910; one in 1933)

*Cecily Hill, Militia Barracks used by 9 Corps, Gloucestershire RVC, 1860 and drill station for 'B' Squadron, RGH, 1914.

*King Street, Bingham Hall built 1908 as a public hall, 'armoury' and rifle range; served as a drill hall for Boys' Brigade and Church Lads' Brigade from 1908; used as Red Cross Auxiliary Hospital in WWI; used as recruiting centre for 7th Battalion, Glosters, 1939. Requisitioned by War Office, 1940, to house Pioneer Corps, NZ Foresters and US 15th Hospital Centre administrative staff. Still in use by Cotswold Rifle Club.

*Somerford Road, Highfield House (1868) ARC, 1960s additions for RGH and 523 LAA Regiment. Currently in use by 'C' Squadron, Royal Wessex Yeomanry.

Coleford

X Town Hall possibly in use in 1910, demolished.
*High Street now cinema, drill hall for 'G' Company, 2nd Battalion, Monmouthshire Regiment (1914) and sometime for 'B' Company.

Dursley

*Masonic Hall/New Inn, Long Street, drill hall for 11 Corps, Gloucestershire RVC, 1860.
X Boulton Lane, chapel of 1823 acquired 1897 as drill hall for 'G' Coy, 5th Battalion in 1914 and shown on 1903 map. In 1947, drill station for 5th Battalion, Glosters.
X Kingshill Lane TAC, 1960s, since rebuilt as Rednock School.

Gloucester

(Three in 1910; three in 1933)
X Militia Barracks, Brunswick Road, HQ and 'A' Squadron, RGH; 3rd Gloucestershire Battery, 1st South Midland Brigade, RFA; HQ + 'A' & 'B' Companies, 5th Battalion, Glosters, 1914.
X Brunswick Road, St Michael's Square roller-skating rink opened 1910, became a drill hall until demolished in 1934; (both dates referenced in *Gloucester Journal*).
*Eastern Avenue, ARC, 1950; HQ 72 AA Brigade 1947; RGH and 5th Battalion, Glosters; in use by 'A' Coy, 6th Battalion, The Rifles, 2021
*Painswick Road; plans in existence 1933, in use 1938 by 5th Battalion, Glosters; in 1947, HQ 72 AA brigade; 498 HAA Regiment; plans for alterations to building 1970, (K/802/130 Gloucestershire Record Office). Now Chequers Bridge Centre, Coney Hill Road.
*Brockworth, Clyde Crescent, cadet centre.
*31 Commercial Road, Custom House, HQ Gloucestershire, Berkshire & Wiltshire Regiment, post-1979.

Minchinhampton

*High Street, possibly Hotel in 1903, drill station for 523 LAA Regiment, 1947; now Cotswold Club.

Moreton-in-Marsh

(One in 1910)
X Station Road, drill hall, demolished *c*. 1960s for Jameson Court housing; drill station for 'H' Coy, 5th Battalion, Glosters, 1914; used by HAA regiment in WWII; drill station for 523 LAA Regiment (TA) in 1947.

Nailsworth

*21–5 Fountain Street, converted Victorian warehouses used by HG and TA; detachment of 523 LAA Regiment in 1947.
*George Street, former Antiques Shop, used by TA pre-Second World War + HG.

Newnham-on-Severn

*Independent Chapel, built 1826, in Littledean (now Dean) Road, drill hall for 12th Corps Gloucestershire RVC from 1860 until 1880, then 'H' Coy, 2nd Volunteer Battalion, Glosters after 1883 and one battery of the Gloucestershire Artillery Volunteer Corps from 1860–1908, becoming The Armoury which, after time as a chapel and a studio, is now the village hall.
*A warehouse of the 1760s on the wharf was adopted as a new drill hall on Back Lane. Both buildings are marked as drill halls on 1903 map but neither of them in use by 1914.

Stow-on-the-Wold

(One in 1910)
The 15th Corps, Glos. RVC constituted a full company in 1868, using the Maugersbury rifle range. Old GS, by 1889 the Masonic Hall, served from 1860 as the Armoury. 'I' Coy, 2nd VB in 1883–1908, with CO, another officer, C/Sgt and drill instructor resident in the town. Drill station for 'E' Coy, 4th Battalion, Oxfordshire & Buckinghamshire Light Infantry based at Chipping Norton in 1914, possibly using St Edward's Hall.

Stroud

(Two in 1910; one in 1933)
X Slad Road drill hall, 'B' Squadron, RGH; (Maj. Playne C/O, Maj. Elwes 2-i/c, and Squadron Sgt-Maj. Swaby, drill instructor in 1914).
X Bath Street drill hall: 'C' Coy, 5th Battalion, Glosters (Lt-Col. Little was CO and C/Sgt Thomas Dennis, drill instructor, in 1914).
X Merrywalks Bowling Green drill hall in use 1902–47+. Bowling Green

belonged to the Golden Heart Inn and became drill ground for RVC who built an armoury and drill hall there. (Capt. W. Davies, CO; T Partidge, surgeon; and Sgt-Maj. Claridge); 'C' Coy, 2nd VB, Glosters. Noted 1885 and 1897 'E' Troop RGH (Maj. Dorington of Lypiatt Park CO); used by HG and demolished for bus station *c.* 1970.

X George Street, Wallbridge depot for Gloucestershire & Worcestershire Brigade Company, South Midland Divisional Transport & Supply Column, ASC (Capt. Ball and G. Hill CSM). In 1947, 1567 Company, RASC.

*38 Bowbridge Lane, Field House, TAC into 1990s; currently converted to apartments but still listed as a recruiting office.

Horns Farm rifle range 1903.

Tetbury

Drill Instructor to 'P' Troop, Badminton Squadron, RGH, lived in Cottons Lane in 1897.

*Malthouse near church in use until 1919 then shooting club; in use as drill station for 3rd Wessex Brigade RFA Ammunition Column, 1914; used by US Army pre-D-Day.

*Royal Oak PH, skittle alley in use by 'B' squadron RGH and detachment 523 LAA Regiment RA TA, in 1960s

Tewkesbury

*Red Lane/Back of Avon; drill hall, 'D' Coy, 2 VB; after 1908, 'D' Coy, 5th Battalion, Glosters. From 1920, 12 Platoon, 'C' Coy, 5th Battalion, Glosters. Currently ACF centre.

Tockington

(One in 1910)

Possibly the Parish Hall marked on the 1913 OS Map.

Winchcombe

(One in 1910)

*Abbey Terrace, Assembly Rooms/Dents Almshouses built in 1865 and served as a hospital 1914–18; also used as working men's club and drill station for 'A' Squadron RGH, 1914.

Wotton-under-Edge

In 1914, drill station for 'D' Coy, 5th Battalion, Glosters. In 1947, drill station for 5th Battalion, Glosters. ACF hut at *Old Grammar School (2021).

APPENDIX VI

Barracks and Camps in the Second World War

Backwell:	US army camp
Beachley:	Army school, currently barracks for 1st Battalion, The Rifles
Bedminster:	US army camp
Beggars Bush:	Airstrip serving nearby US Field Artillery HQs
Blakehill Farm airfield:	Wing HQ, Glider Pilot Regiment
Bristol, Clifton College:	RASC OCTU then HQ 1st US Army
Bristol, Eastville Park:	Tented camp for 5th Battalion, Glosters, 1939–40
Bristol, Horfield Barracks:	Glosters' depot from 1870s–1940
Bristol, Muller Road Orphanage & West of England College of Art:	US billets
Bristol, Muller Road Orphanage:	War Office Holding Centre
Bristol, Old Market Street Drill Hall:	US billets
Bristol, Shirehampton golf course:	Sea Mills, US billets
Bristol, Shirehampton High Street:	Barn as Bristol Garrison HQ, Second World War
Cheltenham Racecourse:	Reinforcement Camp for 164 Brigade, 55 Infantry Division
Cheltenham Racecourse:	US Army combat and medical units
Cirencester Corn Hall and Bingham Library:	Requisitioned by War Office, 1940
Cirencester Militia Barracks:	Built 1857 for North Gloucestershire Militia
Cirencester Park:	Combat Command and three armoured infantry battalions of 5th US Armoured Division, 1944

Cross Hands:	See Springhill Lodges
Daglingworth:	US military hospital & transit camp, then Polish resettlement camp
Daylesford House:	RASC and Durham Light Infantry units
Failand:	US army camp
Fairford Park:	Ex-US 32 Station Hospital became Polish Resettlement Corps camp
Flax Bourton (now Somerset):	GHQ Liaison Regiment (PHANTOM); then US army camp
Gloucester Militia Barracks:	Built 1854–56 for Royal South Gloucestershire Militia
Gloucester, Reservoir Militia Camp:	Built 1939; Glosters' Regimental Depot moved there in 1940 to establish No. 15 ITC which moved to Colchester in 1942; a depot party remained at Reservoir Camp throughout the war, and the 2nd Battalion, Glosters, was in residence in 1944
Gloucester, Robinswood Barracks, Tredworth:	Built 1954, occupied by Royal Signals and RASC; closed and demolished late 1960s
Highnam Court:	Proposed school of military engineering and bridge building 1947
Innsworth:	Imjin Barracks taken over from RAF, 2008; base for NATO Allied Rapid Reaction Corps, and MOD Medal Office
Lechlade:	RHQ GHQ Liaison Regiment (PHANTOM) then to Twickenham
Northwick Park:	US army camp/hospital
Shirehampton:	US army camp
Springhill Lodges:	Militia Camp; 25th Armoured Engineer Battalion US Army, then POW Camp 185 and Polish DPs
Stow-on-the-Wold, Maughersbury Manor:	Base for 22nd Dragoons (flail tanks)
Swindon Hall:	Overflow for 164 Brigade reinforcement camp on Cheltenham Racecourse
Tewkesbury, Old Willow Café (RHQ),	Hop Pole Hotel (officers' mess), Abbey Mill, Red Lane drill hall, barns, etc. and Shakespeare Tavern warehouse (ORs' billets) for 65th (Norfolk Yeomanry) A/T Regiment, 1939
Toddington, Naunton Farm:	Ordnance park pre-D-Day
Ullenwood:	Army hutted camp and US hospital
Westbury on Trym, Badminton School:	HQ 66 (Glosters) S/L Regiment RA

APPENDIX VII

Locations of Volunteer Units

Troops of Yeomanry Cavalry in Gloucestershire and Bristol, 1795

Bristol; Cheltenham; Stow; Dursley; Gloucester; Henbury; Longtree, Bisley & Whitstone; Minchinhampton; and Wotton.

Royal Gloucestershire Hussars Yeomanry, 1891

Troops at Dodington, Cheltenham, Berkeley, Badminton, Chepstow, Monmouth and Tewkesbury.

Royal Gloucestershire Hussars Yeomanry, 1901

'A' Squadron:	Gloucester & Cheltenham Troops, HQ Gloucester
'B' Squadron:	Badminton and Berkeley Troops, HQ Badminton
'C' Squadron:	Chepstow and Monmouth Troops, HQ Newport
'D' Squadron:	Doddington and Bristol Troops, HQ Bristol

Royal Gloucestershire Hussars Yeomanry, 1908

HQ and 'A' Squadron:	Gloucester with Tewkesbury, Ledbury, Cheltenham and Gloucester Troops
'B' Squadron:	Stroud with Cirencester, Cotswold (Stow), Tetbury and Berkeley Troops

'C' Squadron:	Newport with Cardiff, Abergavenny, Newport and Chepstow Troops
'D' Squadron:	Bristol with Old Down and three Bristol Troops

Gloucestershire Artillery Volunteer Corps, 1885

HQ and Nos 1–6 Batteries:		Whiteladies Artillery Ground, Bristol	
No. 7 Battery:	Newnham-upon-Severn	No. 8 Battery:	Gloucester
Nos 9 & 12 Batteries:	Clevedon (Somerset)	No. 10 Battery:	Portishead (Somerset)
No. 11 Battery:	Weston-super-Mare (Somerset)	No. 13 Battery:	Bedminster

Cadet Corps, 1876–1920

Bristol University Senior OTC, 1910 (one infantry company)
Brimscombe Polytechnic Cadet Corps, 1916, affiliated to 5th Battalion, Glosters
Bristol, 26th Company, Boys' Brigade, 1917
Bristol, Brecknell, Munro & Rodgers Cadet Engineer Field Coy, 1918, affiliated to V Bn
Bristol, 1st & 2nd Cadet Battalions, Church Lads' Brigade, 1920
Bristol, Colston's School Cadet Corps, 1915 (four coys.) affiliated to 4th Battalion, Glosters
Bristol GS Cadet Corps, 1900; Junior OTC, 1908 (one infantry company)
Bristol, Kingswood Training School Cadet Corps, 1917, affiliated 2nd Bn Glos. VR
Bristol, Lord Mayor's Own Cadet Corps, 1918, affiliated to 1st Bn City of Bristol VR
Bristol, Merchant Venturers' School Cadet Corps, 1915, affiliated 4Bn Glosters
Bristol, Merrywood School Cadet Unit, 1915, affiliated to 4th Battalion, Glosters
Campden GS Cadet Corps, 1915, affiliated to 5th Battalion, Glosters
Cheltenham Cadet Corps, 1917, affiliated to 3rd Battalion, Gloucestershire VR
Cheltenham College Cadet Corps, 1883: Junior OTC, 1908 (one infantry company)
Cheltenham, Dean Close School Junior OTC, 1909 (one infantry company)
Cheltenham GS (Leslie Young) Cadet Corps, 1916, affiliated to 5th Battalion, Glosters
Cheltenham, St Paul's Cadet Corps, 1917, affiliated to 3rd Battalion, Gloucestershire VR
(Cirencester GS: cadet company 1896, disbanded 1897)
Cirencester, Royal Agricultural College Senior OTC, 1911 (one infantry company)

Cirencester 1st Cadet Company, Boys' Brigade, 1919
Clifton College Cadet Company, 1876; Junior OTC, 1908 (four engineer companies)
Fairfield Higher Grade School, Montpelier Junior OTC, 1915 (one infantry company)
Gloucestershire Regiment, 1st & 2nd Cadet Companies, 1912, affiliated 6th Bn Glosters
Gloucester, 1st & 2nd Battalions, Church Lads' Brigade, 1911
(Gloucester County School, Hempstead: cadet company 1889, disbanded 1891)
Gloucester, Crypt GS Cadet Corps, 1917, affiliated to 5th Battalion, Glosters
Gloucester, Rich School Cadet Corps, 1917, affiliated to 5th Battalion, Glosters
Gloucester, St George's School Cadet Corps, 1917, affiliated to 5th Battalion, Glosters
Gloucester, Wagon Works Cadet Corps, 1918, affiliated to 3rd Battalion, Gloucestershire VR
Stonehouse, Wycliffe College Cadet Corps, 1915, affiliated to 5th Battalion, Glosters
Stroud Cadet Corps, 1918, affiliated to 4th Battalion, Glosters
Stroud, Marling School Cadet Corps, 1916, affiliated to 5th Battalion, Glosters

Volunteer Training Corps (First World War)

City of Bristol

VTC Contingents: Bristol VR /Coliseum Volunteer Corps; Bristol Rugby Volunteer Corps; Bristol Savages VTC; Churchdown, Coalpit Heath & Frampton, Frampton Cotterell, Frenchay, Longhope, Mangotsfield

County of Bristol Volunteer Regiment

HQ National Reserve, 7, College Green; Adjutant's HQ 36, Queen Square; RHQ County Rugby Ground; from 1917 St Michaels Hill Drill Hall

1st Battalion, City of Bristol VR
2nd Battalion, City of Bristol VR
3rd (Athletes) Battalion, Bristol VR, from 1916
Brecknell, Munro & Rogers Cadet Engineer Field Coy and Kingswood Training School Cadet Corps affiliated to 2nd Battalion, City of Bristol VR
Lord Mayor's Own (Bristol) Cadet Corps affiliated to1st City of Bristol VR in 1918

Gloucestershire

VTC Contingents: Berkeley, Birdlip, Calcot, Cheltenham, Churchdown, Cinderford, Cirencester, Cranham, Dursley, Fairford, Forest, Gloucester,

Highnam, Hucclecote & Barnwood, Kingham, Lydney, Nailsworth, Newnham & Arlingham, Ruardean, Staverton, Stroud, Tetbury, Tewkesbury, Twyning, Upton St Leonards

Gloucestershire Volunteer Regiment, June 1915, HQ & drill-ground at Kingsholm, with Forest of Dean, Gloucester, and Cheltenham battalions.

Gloucestershire Volunteer Regiment established with four battalions, 1916, HQ Queen Street, Gloucester.

Gloucestershire Volunteer Regiment, reorganised to include Bristol VR, 1917:

1st (Forest of Dean) Battalion: Newnham, Cinderford ('D'Coy)

2nd (Bristol) Battalion

3rd (Cheltenham & Gloucester) Battalion: HQ & drill-ground Kingsholm; Gloucester ('A'& 'B'Coys), Cheltenham ('C'Coy); Tewkesbury ('D'Coy); Bishops Cleeve, Prestbury and Shurdington

Wagon Works Cadet Corps, Gloucester, and St Paul's Cheltenham Cadet Corps both affiliated to 3rd Battalion, Gloucestershire VR

4th (Berkeley Vale/Stroud) Battalion: Stroud ('A'Coy); Nailsworth Coy; Wotton-under-Edge, Oldbury, Dursley, Chipping Sodbury, Tetbury; Berkeley, Thornbury ('C' Coy), Stow

Stroud Cadet Corps affiliated to 4th Battalion, Gloucestershire VR

Brimscombe Polytechnic Cadets

Following 'assimilation' of City of Bristol battalions:

5th (City of Bristol) Battalion: HQ Clifton Drill Hall, Queens Road

6th (Bristol) Battalion

APPENDIX VIII

Second World War Anti-Invasion Defences

Alderton, DFW3/22 pillbox, SP006333
Bristol, Gypsy Lane, square pillbox, ST612806
Brockworth/Bondend/Nut Hill, pentagonal pillbox with LAA position, SO875150
Cheltenham, Southam, pillbox reduced to two walls, SO974257
Cirencester, re-sited Norcon, 6 loops + entry, SP012012
Cirencester, pairs of loopholes in walls at SU028020 and 019018
Greet, pillbox clad in stone in angle of B4078 and Smithy Lane SP023303
Highnam/Over, DFW3/22 pillbox, SO806195
Maisiemore, DFW3/22 pillbox, SO807218
Mickleton, loopholes in wall at SP162436
Tetbury, Bath Road bridge, formerly 'hairpin' A/T rail sockets
Tetbury, New Church Street, east side, 4 'hairpin' A/T rail sockets, ST888932
Tetbury, New Church Street, west side, 5 'hairpin' A/T rail sockets, ST888932
Tetbury, Gumstool Hill, west side, 4 'hairpin' A/T rail sockets, ST892932
Winchcombe, Langley Road, Langley Hill, pillbox clad in stone, SP006282
Winchcombe, Cotswold Way/Corndean Lane, pillbox clad in stone, SP019270
Winchcombe, concrete sandbag pillbox, SP013274

Airfield Defences

Aston Down ruined DFW3/22 pillbox, SO914019
Aston Down, hexagonal AM type pillbox, ST912998
Chedworth, Battle HQ, SO043125
Little Rissington, rectangular pillbox, SP205188

Little Rissington, rectangular pillbox, rear entrance and LAA position, SP201189
Little Rissington, pillboxes at SP205182, 218193 and 207189, all destroyed
Moreton in Marsh, possible defence work, SP215324
Moreton-in-Marsh, pillbox in top cemetery of Fire Service College
South Cerney, DFW3/22 pillbox variant, SU054983
Southrop, two Norcon pillboxes joined together, SP176036
Southrop, Norcon pillbox with three loops + concrete lid, SP175036
Southrop, two Norcon pillboxes joined together, SP183037, destroyed
Southrop, Norcon pillbox with three loops + concrete lid, SP184037, destroyed
Staverton, pentagonal pillbox with LAA position, SO875226
Staverton, pentagonal pillbox with LAA position, SO892221
Staverton, Bamfurlong Lane, pentagonal pillbox with LAA position, SO893221
Stoke Orchard Battle HQ SO934278
Stoke Orchard Seagull trench SO933279
RAF Windrush octagonal pillbox, AM variant of DFW3/27 design, SP182119

Wye Valley Defences

Bigsweir Bridge, DFW3/22 pillbox on junction, SO540051
Bigsweir Bridge, section post above bridge, SO543052
Bigsweir Bridge, rectangular pillbox above and to east of bridge, SO543054
Bigsweir Bridge, rectangular pillbox above and to east of bridge, SO543053 NB section post on the Welsh bank of the river behind the Toll Cottage, SO538051
Snipehill/Sedbury, section post overlooking bridge, ST552947

Severn Defences, North/West Bank

Beachley Point, DFW3/26 pillbox with prefabricated concrete panels, ST547903
Beachley Point, DFW3/26 pillbox with prefabricated concrete panels, ST549903
Beachley Point, hexagonal pillbox below PH, ST553907
Lydney Harbour, octagonal pillbox with prefabricated concrete panels, SO643013
Woolaston, rectangular pillbox of concrete blocks, 3 loops, ST602991
Woolaston, rectangular pillbox of concrete blocks, 3 loops, ST599989
Woolaston, rectangular pillbox of concrete blocks, 3 loops, ST596987

Severn Defences, South/East Bank

Arlingham, on riverbank, semi buried, SO703126
Arlingham, on riverbank, sunken, rectangular pillbox, 3 loops, SO696102
Arlingham, on riverbank, sunken, rectangular pillbox, SO694105
Arlingham, pillbox built on grounded barge, SO696115 destroyed
Purton, rectangular pillbox, 1 loop per side + entry, SO691046
Slimbridge, rectangular pillbox behind sea wall, SO720054
Slimbridge, rectangular pillbox behind sea wall, SO712046
Slimbridge, DFW3/?27, 8-sided with varied loops, levelled, SO708042
Slimbridge, rectangular pillbox, three loops, SO719050

GHQ Line Red: Type DFW3/22 (Shellproof) Pillboxes

Buscot SU225983, SU227983, SU229983, SU235982 and SU239983
Lechlade, St John's Bridge SU221993

Green Line Defences: Long Newnton to Avening

(Numbers relate to Major Green's survey)
Long Newnton, Type DFW3/26 Green Line pillbox, ST918915 (12) NB similar (13) over county boundary.
Long Newnton, Type DFW3/24S Green Line pillbox, ST919920 (11)
Long Newnton, Type DFW3/24S Green Line pillbox, ST918927 (10)
Long Newnton, Type DFW3/24S Green Line pillbox, ST915932 (9)
Long Newnton, Type DFW3/24S Green Line pillbox, ST912934 (8)
Long Newnton, Type DFW3/24S Green Line pillbox, ST909935 (7)
Tetbury, A/T obstacles by railway, ST905941 (6)
Tetbury, Summerwell Farm, Type DFW3/24S Green Line pillbox, ST902961 (5)
Tetbury, Star Farm, Type DFW3/24S Green Line pillbox, ST898965 (4)
Tetbury, Star Farm, Type DFW3/24S Green Line pillbox, ST889971 (3)
Avening, Type DFW3/24S Green Line pillbox & A/T obstacles, ST887978 (1 & 2)

GHQ Green Line Defences: Upper Framilode to Stonehouse: Stroudwater Canal and River Frome

Upper Framilode, DFW3/24S pillbox, SO752101
Whitminster, Fromebridge Mill, DFW3/24S pillbox, SO772075
Whitminster, near A38, DFW3/24S pillbox, SO711073

Frampton on Severn, Frome Bridge, DFW3/24S sunken pillbox, SO770070
Eastington, Green Line DFW3/24S pillbox, SO783061
Eastington, Green Line DFW3/24S pillbox, SO781057
Stonehouse, Bonds Mill, DFW3/24S pillbox with added brick porch, SO793050
Stonehouse, Bonds Mill, two-storey octagonal pillbox + LAA position, SO793053

GHQ Green Line/Sharpness Canal Defences

Saul, DFW3/26 pillbox with brick porch, SO757094
Whitminster, Stonepitts Bridge, DFW3/26 pillbox, SO766084

APPENDIX IX

Home Guard Units

32,000 men in nineteen battalions: 1st (Cheltenham/Tewkesbury) Battalion; 2nd (Campden) Battalion; 3rd (Cirencester) Battalion; 4th (Forest of Dean) Battalion; 5th (Gloucester) Battalion; 6th (South Gloucester) Battalion; 7th (Stroud) Battalion; 8th (Dursley) Battalion; 9–14th, 16th & 18th (City of Bristol) Battalions, in Bristol City Sector, South-west District Southern Command; 15th (GPO, Bristol) Battalion; 17th (Wye Valley) Battalion; 19th (Churchdown) Battalion. 1–8th, 17th and 19th Battalions in South Midland District, Southern Command.

The following were all under local battalion control: 17 AA Regiment, Home Guard (Bristol); 141 (101) Gloucestershire Home Guard AA Rocket Battery (Bristol); 182 (102) Gloucestershire Home Guard AA Rocket Battery (Bristol); 210 (103) Gloucestershire Home Guard AA Rocket Battery (Bristol); 215 (104) Gloucestershire Home Guard AA Rocket Battery (Abbots Leigh, Bristol); 71 (Gloucestershire Home Guard HAA Battery (Bristol); 143 (105) Gloucestershire Home Guard AA Rocket Battery (Gloucester); Gloucestershire Home Guard (Independent) LAA Troop (Gloucester).

Five LAA batteries under 15th and 18th Battalions, Gloucestershire Home Guard defending Filton and further batteries at Hucclecote, Staverton (Rotol), Stonehouse (Hoffman), and two at Cheltenham (Smiths and Dowty).

Some Premises used by Gloucestershire Home Guard in the Second World War

Berkeley, Swan Inn, Long-room: lecture room HG training
Bristol, Filton Airport: HQs 13th and 18th Battalions

Bristol, Filton, Station Road, AM Depot: HQ 'D' Company, 9th Battalion
Bristol, Hillside: HQ 15th Battalion
Bristol, Horfield, Bishopston parish hall: HQ 'B' Company, 9th Battalion
Bristol, Kellaway Avenue, GS Grounds: HQ 9th Battalion and 'C' Company
Bristol, Kingsdown, St Matthews parish hall: HQ 'A' Company, 9th Battalion
Bristol, Old Market Drill Hall: HQ 16th Battalion
Bristol, Packers Factory: HQ 12th Battalion
Bristol, 32 Park Row: Bristol Fort and sub-area HQ
Bristol, Redcliffe Parade: HQ 10th Battalion
Bristol, St Andrews, 99 Somerville Road: HQ 9th Battalion, from 1942
Bristol, Upper Belgrave Road: HQ 11th Battalion
Cheltenham, Lypiatt Terrace: HQ 1st Battalion
Cheltenham, North Street drill hall: HQ Company 1st Battalion
Cirencester, The Abbey: HQ 3rd Battalion
Cirencester, Cecily Hill, Militia Barracks: HQ 'C' Coy, 3rd Battalion
Dursley Court: HQ 8th Battalion
Falfield: Anti-Gas School
Gloucester, Painswick Road, drill hall: HQ 5th Battalion
Hartpury Parish Hall: 21 Platoon, 'E' Coy, 5th Battalion
Horfield, Wellington Garage: ZAA training
Stonehouse, Stanley Mill: HQ 'E' Coy, 8th Battalion
Stroud, London Road, Western House: HQ 7th Battalion
Tewkesbury, Red Lane drill hall: HQ 'A' Company, 1st Battalion
Tewkesbury, Old Hospital: HQ 'E' Company, 1st Battalion

APPENDIX X

Air Defence in the Second World War

Anti-Aircraft Batteries

Bristol's HAA Batteries recorded in June 1941, with renumberings in brackets: B1(3) Gordano; B2(1) Portishead; B3(7) Rockingham Farm; B4(5) Cribbs; B5(11) Brickfields; B6(13) Purdown; B7(17) Whitchurch; B8(19) Reservoir; B9 Avonmouth; B10(9) Almondsbury; B11 Hanbrook; B12(15) Hanham; B14 Blackboy; B15; B17(14) Rodway; Lodge Farm (2); St Georges Wharf (4); Westbury (6); Pilning (8); Earthcote (10); Henfield (12) unarmed; Keynsham (16); Chew (18); Backwell (20).

Gloucester's HAA Batteries recorded as armed in June 1942: A12 Haydons Elm; A13 Dean Farm; A14 Belmont; A15 Parton; B3 Buckholt; the rest, A1-11, B1 and B5 were all unarmed.

Bristol ZAA Batteries armed in June 1942: Z1 Easton in Gordano; Z2 Bishopsworth; Z5 Brislington

Military Bombing Decoys: RAF and Army

Ashton Keynes (SU028927) Somerset: Q decoy for RAF South Cerney
Horsley (ST865963): Q decoy for RAF Aston Down
Farmington (SP146171): Q decoy for RAF Little Rissington
Patchway (ST630819) Q/QF decoy for Filton airfield (RAF & BAC)
Haresfield and Longney (SO774120) QF decoys for RAF Quedgeley
Leightertown (ST829924) decoy site on WWI airfield
Twigworth (SO841243) QF for RAF Whitchurch, Gloucester
Fiddington (SO943306): QF decoy for Ashchurch RASC & SOS vehicle depot

Civil Bombing Decoys: C and P Series and Starfish

Bristol: Sheepway QF (ST497779); Long Ashton QL; Chew Magna QL (ST575648); Stockwood QL (ST629679); Severn Beach QF (ST542836); Lawrence Weston QF (ST538787); Patchway QL/QF (ST630819); Yeomouth QL (ST368670); Downside QL (ST478659); Kenn Moor QL (ST436685); Cheddar SF decoy (ST466557)

Bristol (Mendips) six sites at Burrington QL/QF (ST465557-497570)

Gloucester: Longney QL (SO774120); Standish QL (SO832085); Brimpsfield QL (SO927125); Coberley (SO972168)

Severn Tunnel, Redwick QL/QF

Avonmouth, Sheepway and Severn Beach, Oil QF decoys (P Series)

Royal Observer Corps Posts

NB: there are some differences in this list from those compiled by Derek Wood in 1976, and by Dr Colin Dobinson in 2000 for the Defence of Britain project. This list has been compiled using all available information including Subterranea Britannica's field reports and personal observations.

HQs and control centres

Interwar years: No. 23 Group, Bristol: HQ Little King Street, rear of Public Library, 1937. No. 24 Group, Gloucester, HQ GPO Exchange, George Street 1937.

Second World War: No. 23 Group, Bristol; HQ Worcester Terrace, 1943–5. No. 24 Group, Gloucester. HQ GPO Exchange, George Street to 1942; then Northgate Mansions.

Cold War: No. 12 Group, Bristol, 1953–1991; HQ Kings Square Avenue, 1947–58; then move to Protected Accommodation in former AAOR Lansdown in Bath, reporting to Centralised Combined Filter Plot at RAF Box in Southern RAF Sector. No. 24 Group, Gloucester. HQ Northgate Mansions 1945–1953.

ROC posts:

	1937–8	WW2	1945–59	1960-68 u/g	1968-91 u/g	Comments
Alderton SP008327		X				Granite, 1942
Now Evesham SP019408				X	X	Alderton resite
Almondsbury ST609842		X				
Move to ST596834	X	X				Almondsbury resite
Now Thornbury ST647888				X		Almondsbury resite
Andoversford SP028195		X	X			Granite 1942
Move to SP032208		X				Andoversford resite
Move to SP049219			X		X	Andoversford resite
Avonmouth ST518782		X	X			
Now Hallen ST520807				X		Avonmouth resite
Move to ST553803				X	X	Hallen resite
Blakeney SO668061	X	X		X	X	
Broadway Tower SP113362	X	X		X	X	Granite 1943
Brockworth SO875150			X	X		on Nut Hill
Now Upton St Leonards SO875150				X	X	renamed Brockworth
Cleeve Hill SO987283		X				Granite 1942
Move to SO987278		X				Cleeve Hill resite
Move to SO983283			X			Cleeve Hill resite
Dursley ST754984	X					
Move to ST861989		X				Dursley resite
Move to ST761989			X			Dursley resite
Move to ST733996				X	X	Dursley resite
Hill ST654969		X				
Move to ST654953			X			Hill resite

Move to ST643953				X		Hill resite
Hucclecote SO865170			X			see Brockworth
Leighterton ST823920		X	X	X		
Move to ST830893				X	X	Leighterton resite
Little Sodbury ST759825			X	X	X	
Marshfield ST768725	X	X	X	X	X	
Minsterworth SO756168		X	X			
Now Westbury-on-Severn SO731145				X		Minsterworth resite
Move to SO731145				X		Westbury resite
North Cerney SP031081		X	X	X		
Move to SO029081				X		North Cerney resite
Northleach SP096146		X	X	X		
Poulton SP112027	X	X				
Move to SP101005			X			Poulton resite
Move to SP100020				X		Poulton resite
Rodmarton ST937976		X	X			
Now Kemble, ST937985				X		Rodmarton resite
Move to ST967963				X	X	Kemble resite
St Briavels SO559041	X	X	X	X	X	
Stoke Orchard (airfield) SO934278				X	X	
Stow-on-the-Wold SP193267	X	X	X	X	X	Granite 1943
Stonehouse SO806051		X	X			Wycliffe College.
Now Middle Lypiatt SO876049 then to SO829030				X		Stonehouse resites
Now Stroud SO831086				X	X	Middle Lyppiatt resite
Tirley SO802279		X	X	X	X	

APPENDIX XI

Intelligence and Communications

Second World War

Birdlip: Emergency transmitter produced by RN signals school installed by RAF No. 80 (Signals) Wing

Boddington: military telephone exchange run by ATS

Bristol: Admiralty Signals Establishment, 1940

Bristol, Stockland: Radio Security Service/MI6 DF Station

Down Ampney, Williamstrip Camp: base for SAS Brigade, May 1944

Fairford, Hatherop Castle: SOE Operational School STS 45; Force 12 (enemy aliens) and training centre for Danish 'Jedburghs'

Fairford, The Cage: secure transit camp for SAS personnel

Forthampton House, Tewkesbury, SOE Operational School STS 49, formerly STS 45a, used mainly by the Polish Country Section

Leckhampton Hill: HF/DF station

North Cerney: MI6 Section VIII radio transmitter serving Whaddon/Windy Ridge, PWE

Cold War

Benhall: GCHQ Intercept School within main establishment one of the two sites at Cheltenham; later transferred to The Royal Signals Centre in Loughborough

Boddington: No. 9 RAF Signals Unit, messaging relay centre, under Support Command Signals HQ, 1978–2008

Bristol, RAF (then HMS) Pucklechurch (ex-Balloon depot): joint services technical training depot

Bristol, Purdown: Backbone microwave tower
Cheltenham, Oakley Farm: GCHQ, 1952
Cheltenham, Benhall: GCHQ research departments and experimental radio station from Knockholt (Kent), 1952; new building 2003 (The Doughnut)
Wotton-under Edge: Backbone microwave tower

APPENDIX XII

Munitions Production, Depots, and Logistics

First World War

Avonmouth: ASC MT Mobilisation Depot, 1914
Avonmouth: ASC Tractor Depot, 1915
Avonmouth Docks: used by ASC for MT and petrol shipments
Beachley Peninsula: National Shipyard, 1916
Brislington: footwear factories
Bristol, Ashton Halt: assembly point for horses and mules
Bristol, Fishponds: Brazil Straker works building aero engines
Bristol, Gloucester Road, Bristol North Baths, munitions training centre, 1917
Bristol, Netham Works: United Alkali Co., chemical works
Bristol, Park Row rink: Parnall & Sons building Short Bros aircraft on licence
Bristol, St Philips: John Lysaght Ltd manufacturing barbed wire
Bristol, St Philips, Victoria Road: shell-case collection, assembly, and acceptance centre
Bristol, St Philips, Victoria Road: Castle Iron Works making 18-pounder shells
Bristol: Douglas Motorcycle Company made 25,000 motorcycles for armed forces
Brockworth/Hucclecote: flight test airfield for HH Martyn & Co. of Cheltenham, sub-contractor for Aircraft Manufacturing Co., Gloucester Aircraft Co. in 1917
Cheltenham Winter Gardens leased by HH Martyn for aircraft production
Chittening St Andrews Road: National Filling Factory No. 23, for mustard gas shells; Holland & Cubitts Grillo Oleum Plant Absorption House
Coleford, Speech House/Monmouth Roads: HM Acetone Factory
Dursley: Listers made shells
Dursley: Mawdsleys making dynamos and motors
Easton, Thrissell Street ex-Baptist chapel: Brecknell, Monro & Rogers, shell cases
Filton: Bristol Aeroplane Co. & RFC Aircraft Acceptance Park, 1917

Filton: footwear factories
Gloucester Wagon Works making shells
Henbury: HM Factory, explosives factory until 1917, unfinished
Quedgeley: National Filling Factory No. 5, explosives factory
Shirehampton: Army Remount Depot, 1915
Slimbridge: HM No. 23 Magazine
Stroud: Daniels making shells
Warmley: Crane & Co. filling hand grenades
Yate: No. 3 (Western) Aircraft Repair Depot, 1916

Interwar Period 1919–39

Aston Down: No. 20 MU (ASU), 1938
Hucclecote: Gloster Aircraft Co, 1926, taken over by Hawker Aircraft Co. Ltd, 1934
Little Rissington: No. 8 MU (ASU), 1938
Tetbury: Army Mobilisation Depot for storage of AA equipments
Yate: Geo. Parnall & Co. 1925; building Percival prototypes 1931; Hendy Aircraft Co. and Nash & Thompson Ltd. amalgamated to form Parnall Aircraft Ltd, 1935

Second World War

Ashchurch, Tewkesbury: RASC and US Army Service of Supply Storage Depots
Aston Down: RAF No. 20 MU (ASU)
Aston Down: RAF No. 424 Advanced Fuel and Ammunition Park, 1944
Berwick Wood: fuel supply junction linking Avonmouth to national fuel ring-main
Bishops Cleeve: Smiths Instruments relocated factory
Bristol University, HH Wills Physics Laboratory: ASE Valve Division from Eastney
Cheltenham: Folland Aircraft Co.
Cheltenham: Gloster Aircraft Co. later moved to Brockworth
Downleaze, Bristol: HQ RAF No 1 Works Area (Field), 1940
Dursley: Boulton Mills dispersal for Parnall's Yate MAP factory
Filton: Bristol Aeroplane & Engine Company
Gloucester Railway Carriage and Wagon Company building tanks and rolling stock
Gloucester: Purgatory Area centre, for storage of uncompleted aircraft
Hallen: Air Ministry AFRD and area HQ for fuel supply
Hucclecote/Brockworth: Gloster/MAP shadow factories opened 1940
Hawthorn, Forest of Dean: Admiralty storage depot
Little Rissington: RAF No. 8 MU
Lydney Harbour: US Service of Supply depot

Lydney Harbour: aircraft factory making parts for gliders
Moreton Valence, Gloucestershire: flight test facility for Glosters
Newland: RAF No. 59 MU, Air Ammunition Park, 1940
Newland: RAF Forward Ammunition Depot, No. 59 MU, 1942–5, to Rhoose, Cardiff
Quedgeley/Hardwick: RAF No. 7 MU with depots on seven sites
Staverton: MAP-sponsored companies carrying out aircraft repairs
Staverton: Rotol Airscrews Ltd, Flight Test Department, 1939
Staverton: Folland Aircraft flight testing, 1940–1
Stoke Orchard: shadow factory run by Glosters assembling Typhoons
Stonehouse, Bond's Mill: Sperry Gyroscope factory
Stonehouse, Oldends Lane: Hoffmans' ball-bearings works
Yate: Parnall Aircraft Ltd; producing Spitfire components, 1939–45

Cold War and Beyond

Ashchurch, Tewkesbury: Army Logistic Storage Depot
Fairford: USAF SAC nuclear weapons storage
Filton: Bristol Aviation: Bloodhound SAM production 1950s–1980s
Moreton Valence: factory & flight testing for Glosters, 1946–1962
Quedgeley: RAF No. 7 MU Depots

APPENDIX XIII

Military and Auxiliary Hospitals

First World War

Badminton: Portcullis House
Bitton: Beach House; Boddington Hospital
Bourton-on-the-Water: Copse Hill
Bristol: Ashton Court; Beaufort War Hospital (Glenside Asylum); Kingsdown: 2nd Southern General Hospital RAMC; Kingswood School; Southmead Workhouse Infirmary; Bishop's Stoke, Bishop's Knoll House; Stoke Park Hospital; Crump Farm, Pucklechurch; Almondsbury Subsidiary Military Hospital; Manor Hospital, Tockington; Bruce Cole Institute, conversion of Elizabeth Shaw chocolate factory, Packers' sports pavilion, in Whitehall, Greenbank
Cheltenham: Racecourse Hospital; New Court; Suffolk Hall; Moorend Park, Charlton Kings; The Abbots; St Martins; Naunton Park Council Schools; St Johns Gloucester Road Council Schools and The Priory
Chipping Campden: Norton Hall
Chipping Sodbury: Harley House, Horton Hall and Hawkesbury
Cirencester: Bingham Hall
Downend: Cleve House
Dumbleton: Village Hall
Gloucester: Hillfield House; The Palace; Kings Weston House; Leckhampton Court
Lydney: Town Hall, secondary school and Redhill
Moreton-in-Marsh: Kitebrook
Nailsworth: Chestnut Hill
St Briavels: The Ghyll House
Stonehouse: Standish House
Stow-on-the-Wold: Abbotswood
Stroud: Trinity Parish Rooms
Tewkesbury: Memorial Hall and Mitton Farm
Westbury-on-Trim: Red Maids' School
Winchcombe: Assembly Rooms.

Interwar Years

Bristol, Stoke Park Hospital: became Burden Neurological Institute, 1939

Second World War

Andoversford, Brockhampton Park: Red + Convalescent Home/Auxiliary Hospital
Bristol, Frenchay: US Army Hospital
Bristol: US Western Base Section General Hospital, 600-bed EMS
Bristol: Army Blood Transfusion Service
Cheltenham: HQ QAIMNS, 1942–5
Cheltenham Racecourse: 16th US Army Station Hospital
Cheltenham: US Southern Base Section General Hospital, 834-bed EMS
Chipping Sodbury, Little Sodbury Manor: Red + Convalescent Home/Auxiliary Hospital
Cirencester Park: 15th US Army Hospital Centre, then Polish Resettlement Corps camp
Cirencester, Cerney House: Red Cross Convalescent Home/Auxiliary Hospital
Coates, Dockem House: Red Cross Convalescent Home/Auxiliary Hospital
Daglingworth: No. 40 US General Hospital
Down Ampney: central air evacuation casualty base, 1944–46
Fairford Park: No. 32 US General Hospital, then Polish Resettlement Corps camp
Falfield: No. 91 US Evacuation Hospital
Gloucester, Hartpury: Red Cross Convalescent Home/Auxiliary Hospital
Innsworth: hutted RAF Station Hospital 1940–48
Minchinhampton, Nos 6 and 28 US Field Hospitals, and No. 32 US General Hospital
Northleach, Stowell Park, No. 160 US General Hospital
Northwick Park: No. 327 US General Hospital, 1943–4, then POW hospital camp and Polish Resettlement Corps camp
Stroud, Edgeworth Manor: Red Cross Convalescent Home/Auxiliary Hospital
Tetbury, Estcourt: Red Cross Convalescent Home/Auxiliary Hospital
Tortworth: Nos 128 & 91 US Evacuation Hospitals 1943–4; Nos 94, 224 & 225 General Hospitals, 1944–5
Ullenwood: No. 62 US General Hospital
Wotton-under-Edge: No. 45 US Evacuation Hospital

Post-Second World War

Little Rissington: USAFE Contingency Hospital, 1982

APPENDIX XIV

POW Camps in the Second World War

Woodfield Farm Camp, Churchdown, No. 1
Sudeley Castle, Winchcombe, No. 37
Coleford, Wynolls Hill, No. 61 (satellite of Leckhampton)
Wapley Camp, Yate, No. 124
Bedminster, Ashton Gate Camp, No. 124/124a
Northwick Park Camp, No. 132
Bourton-on-the-Hill, No. 157
Springhill Lodges Camp, No. 185
Leckhampton Court Camp, No. 263
Northway Camp, Ashchurch, No. 1009
Churcham Camp near Highnam
Lydney, Naas Lane, Naas House Camp
Siddington Hall (until 1948)

Bibliography

Aberg, F., 'Medieval Moated Sites', *CBA Research Report* 17 (York: 1978)
Abraham, B., 'Barnsley Park No.22 SLG', *Airfield Review* 91 (Stockport: ARG, 2001)
Adams, M., *The First Kingdom: Britain in the Age of Arthur* (London: Head of Zeus, 2021)
'Historical Appraisal RAF Quedgeley', *Airfield Research Group Report* 25, (ARG, 2004)
Aston, M., 'The Lost County of Winchcombeshire', *British Archaeology* 92 (York: CBA, 2007)
Atkin, M., 'Gloucester's Norman Castle Rediscovered', *FORTRESS* 9 (Liphook: Beaufort, 1991); 'The Civil War Defences of Gloucester', *FORTRESS* 10 (Liphook: Beaufort, 1991)
Atkin, M., & Laughlin, W., *Gloucester and the Civil War* (Stroud: Alan Sutton, 1992)
Aughton, P., *Bristol: A People's History* (Lancaster: Carnegie Publishing, 2000)
Barratt, J., *Sieges of the English Civil Wars* (Barnsley: Pen & Sword, 2009); *A History of Maritime Forts in the Bristol Channel 1866–1900* (Shirehampton: self-published, 1978 and 1993)
Berry, P., 'OAMCU Gloucester 1940', *Airfield Review* 73 (Ware: ARG, 1996)
www.binghamhall.co.uk/history (Cirencester)
Bristol City Council, *Purdown Heavy Anti-Aircraft (HAA) Battery, Stoke Park Estate, Bristol* (Bristol: Cotswold Archaeology, 2015)
'Bristol During World War Two', humanities.uwe.ac.uk/bhr/Main/ww2/4_13.htm
Casey, D., 'Lydney Castle', *Antiquaries Journal* XI (London: 1931)
Corley, D., 'Fairford: A Base for the Bigger Aircraft from 1944 to 1990', *Airfield Review* 123 (Thetford: update of *Airfield Review* 53, 1990, Airfield Research Group, 2009)
Crowther, S., and Dickson, A., *Severn Estuary Rapid Coastal Zone Assessment Survey* (GCC, EH & SCC, 2008)
Darvill, T., *Prehistoric Gloucestershire* (Stroud: Amberley, 2013)
Davies, J., and Phillips, C., *The Percy Sladen Memorial Fund Excavations at Bury Hill Camp, Winterbourne Down. 1926* (Bristol: University of Bristol Speleological Society Proceedings 3, 1927)
Davis, P., www.gatehouse-gazetteer
Dean Archaeological Group, *Moated Sites Survey* (Coleford: DAG, 1998)

Dickson, J., 'Helping to defend our country and Hoffmann's Shadow Factory', *Stonehouse History Group Journal* 1, 2012

Dixon, P., *Crickley Hill Volume 1: The Hillfort Defences* (Crickley Hill Trust and University of Nottingham, 1994)

Dodd, A., and Moss, P., 'Brimpsfield Castle and the Giffards', *Glevensis* 25 (Gloucester: Gloucestershire Archaeology, 1991)

Ecclestone, M., 'What happened to Harescombe Castle', *Glevensis* 40, (Gloucester: Gloucestershire Archaeology, 2007)

Edwards, B., 'The Slimbridge Munitions Depot', *Gloucestershire Society for Industrial Archaeology Journal* (Gloucester: GSIA, 1995)

Ferris, J., *Behind the Enigma: The Authorised History of GCHQ, Britain's Secret Cyber-Intelligence Agency* (London: Bloomsbury, 2020)

Finney, J., *Middle Iron Age Warfare of the Hillfort Dominated Zone c400BC to c150BC* (unpublished PhD thesis, University of Bournemouth, 2005)

www.flickr.com/photos/13150208@N05/albums. Photos of five Bristol HAA sites: Rockingham Farm; Easton-in-Gordano; Portway, Sheepway; Lodge Farm, Portishead; and Purdown

Fry, J. S., & Sons Ltd., *English City; The Growth and the Future of Bristol* (Bristol, 1945)

Goodall, J., *The English Castle 1066–1650* (Newhaven CT. & London: Yale, 2011)

Grace, P., Weekly War Diary for 1939, extracted from *Wiltshire & Gloucestershire Standard*

Graham Kerr, A., 'Defence of the Wye & Usk Valleys', *Loopholes* 9 (Dover: Pillbox Study Group, 1994)

Green, Maj. M., and Plant, J., 'Bristol Outer Defences on the Cotswold Plateau parts 1 & 2 + Annexe', *Loopholes* 4, 5, and 7 (Dover: Pillbox Study Group, 1993–4)

Green, Maj. M., *WARWALKS Stop Line Green* (Cheltenham: Reardon, 1999)

Guy, N., 'Thornbury Castle: Renaissance Palace or stronghold of a feudal baron?' *Castle Studies Group Journal* 19 (Daventry: CSG, 2006)

Hannan, A., 'Holm Castle, Tewkesbury excavations', *Glevensis* 9 and 10 (Gloucester: Gloucester & District Archaeological Research Group, 1975–77)

Higgins, D., *The History of the Bristol Region in the Roman Period* (Bristol: Bristol Branch, Historical Association, 2005)

Holbrook, N., *The Roman Period* (Gloucester: Cotswold Archaeology, 2011)

Howes, R., 'Gloucester Castle as a Prison', *Glevensis* 43 (Gloucester: Gloucester & District Archaeological Research Group Review, 2010)

Hoyle, J., 'Finding the history hidden in the forest', *British Archaeology* 126 (York: CBA, 2012); 'Research Framework for the Forest of Dean District, Forest of Dean Archaeological Survey Stage 4 Module 3' (Gloucester: GCC, 2017)

Hurst, H., 'The Archaeology of Gloucester Castle: an introduction', *Transactions of Bristol & Gloucestershire Archaeological Society* 102 (Gloucester: BGAS, 1984)

Israel, R., *The Archaeology of the Fortifications constructed in England during the English Civil War (1642–51)*, focus on Bristol and Gloucester (Bristol: unpublished PhD Thesis, University of Bristol, 2020), research-information.bristol.ac.uk

Kelly, M. (ed.), *Bristol and the First World War* (Bristol: Bristol Cultural Development Partnership, 2014), www.bristol2014.com

Kenyon, D., *First World War National Factories* (Cambridge: Historic England online, 2015)

King, A., 'A Royalist Bastion? Evidence from 30 Gloucester Lane, Old Market, Bristol', *Transactions of the Bristol & Gloucestershire Archaeological Society* 128, 2010

MacInnes, C., *Bristol at War* (London: Museum Press, 1962)

Moore-Scott, T., *The Archaeology of Leckhampton Moat*, leckhamptonlhs.weebly.com

Norton, S., *Assessing Iron Age Marsh Forts* (Birmingham: unpublished PhD thesis University of Birmingham, 2019) etheses.bham.ac.uk

Osborne, M., *Sieges & Fortifications of the Civil Wars in Britain* (Leigh-on-Sea: Partizan Press, 2004); *Defending Britain* (Stroud: Tempus, 2004); *Always Ready: The Drill Halls of Britain's Volunteer Forces* (Leigh-on-Sea: Partizan Press, 2006); *Pillboxes in Britain and Ireland* (Stroud: Tempus, 2008); *Grandad's Army: Volunteers Defending the British Isles in the First World War* (Stroud: Fonthill Media, 2020)

Penny, J., 'The Air Defence of the Bristol Area 1937–44' (Bristol: Bristol Branch of the Historical Association, Pamphlet No. 90, 1997); 'South Gloucestershire Compendium 1914 to 1918'

Putley, J., and Webb, A., *Home Front Dean* (Coleford: Dean Archaeological Group, 2004)

Rawes, B., 'A Check List of Castles and other fortified sites of Mediaeval date in Gloucestershire', *Glevensis* 11 (Gloucester: Gloucester & District Archaeological Research Group, 1977); 'A Preliminary Checklist of Moated Sites in Gloucestershire', *Glevensis* 12 (Gloucester: Gloucester & District Archaeological Research Group, 1978)

Reid, B., 'The Minchinhampton Bulwarks: a post-Roman Frontier?' *Glevensis* 32 and 33 (Gloucester: Gloucester & District Archaeological Research Group, 1999 & 2000)

RCHM(E): *Ancient & Historical Monuments in the County of Gloucester: Iron Age & Romano-British Monuments in the Gloucestershire Cotswolds* (London: HMSO, 1976)

Rowe, P., *'We shall defend our island'- Investigating a forgotten militarised landscape* (unpublished PhD thesis, University of Southampton, Faculty of Humanities, Archaeology, 2014)

Russell, J., *The Civil War Defences of Bristol: Their Archaeology & Topography* (Bristol: self-published, 1995)

Salter, M., *The Castles of Gloucestershire & Bristol* (Malvern: Folly Publications, 2002)

Seymour, W., *Battles in Britain 1066–1746* (Ware: Wordsworth, 1997)

Short, N., 'Gipsy Patch Lane Pillbox', *Loopholes* 69 (Burgh: Pillbox Study Group, 2017)

Spry, N., Researching the City of Gloucester, www.glosarch.org.uk/Gloucestercityresearch.html

Strickland, A., *The Stroudwater (Canal) Pillboxes* (Gloucestershire Society for Industrial Archaeology, www.gsia.org.uk, 2007)

Turner, M., *Secret Forest of Dean* (Stroud: Amberley, 2019)

Verey, D., *The Buildings of England, GLOUCESTERSHIRE 1: The Cotswolds; GLOUCESTERSHIRE 2: The Vale and the Forest of Dean* (Harmondsworth: Penguin Books, 1970 & 1999/2000 with Brooks, A.)

Wakefield, K., *Operation Bolero: the Americans in Bristol and the West Country 1942–45* (Manchester: Crecy Books, 1994)

Walker, D., 'Gloucestershire Castles', presidential address in *Transactions of the Bristol & Gloucestershire Archaeological Society* 109 (Gloucester: B&GAS, 1991)

Webb, D., 'Wycliffe at War', *Stonehouse History Group Journal* 1, 2012

Wikipedia lists of Regular Army and Army Reserve units and locations accessed 14.1.2021

Williams, G., 'Cirencester', *Loopholes* 84 (Burgh: Pillbox Study Group, 2022)

Wood, M., *Domesday A Search for the Roots of England* (London: BBC Publications, 1986)

Yeates, S., 'Could the British be older than Britain?', *British Archaeology* 104 (York: CBA, 2009)

Index